Elasticities in International Agricultural Trade

Elasticities in International Agricultural Trade

EDITED BY

Colin A. Carter
and Walter H. Gardiner

Routledge
Taylor & Francis Group
LONDON AND NEW YORK

First published 1988 by Westview Press, Inc.

Published 2018 by Routledge
52 Vanderbilt Avenue, New York, NY 10017
2 Park Square, Milton Park, Abingdon, Oxon OX14 4RN

Routledge is an imprint of the Taylor & Francis Group, an informa business

Library of Congress Cataloging-in-Publication Data
Carter, Colin Andre.
Elasticities in international agriculture trade / edited by Colin Carter and Walter H. Gardiner.
p. cm.—(Westview special studies in international economics and business)
ISBN 0-8133-7563-0
1. Produce trade. 2. International trade. 3. Elasticity (Economics) I. Gardiner, Walter H. II. Title. III. Series.
HD9000.5.C278 1988
382′.41—dc19 88-5781
CIP

ISBN 13: 978-0-367-01208-3 (hbk)

Contents

Preface iv

1. **Issues Associated with Elasticities in International Agricultural Trade** 1

   Walter H. Gardiner and Colin A. Carter

2. **Elasticities in International Trade: Theoretical and Methodological Issues** 17

   Jerry G. Thursby and Marie C. Thursby

3. **Estimating U.S. Agricultural Export Demand Elasticities: Econometric and Economic Issues** 53

   Philip C. Abbott

4. **The Role of Exchange Rates in Trade Models** 87

   John Dutton and Thomas Grennes

5. **Macroeconomic Shocks, Foreign Trade, and Structural Adjustment: A General Equilibrium Analysis of the U.S. Economy, 1982-1986** 137

   Irma Adelman and Sherman Robinson

6. Price and Exchange Rate Transmission Revisited: The Latin America Case 163

Christine Bolling

7. Market Share Models and the Elasticity of Demand for U.S. Agricultural Exports 195

David Blandford

8. Export Demand Elasticity in the World Market for Beef 225

Ellen W. Goddard

9. Imperfect Price Transmission and Implied Trade Elasticities in a Multi-commodity World 255

Rod Tyers and Kym Anderson

10. Discussion and Concluding Comments 297

William H. Meyers, Andrew Schmitz, Robert Thompson, and Alex F. McCalla

Contributors 303
Index.................................305

Preface

This collection of papers was presented at the 1987 IATRC Symposium on Elasticities in International Agricultural Trade in Dearborn, Michigan. It is the outgrowth of renewed interest by researchers, policymakers, traders, and others in quantifying those factors that affect international trade of agricultural products. The papers presented at this conference address a number of the theoretical and empirical issues related to the estimation, interpretation, and application of elasticities in international agricultural trade.

Following the introductory chapter by Colin A. Carter and Walter H. Gardiner, the first paper, by Jerry Thursby and Marie Thursby, surveys the recent research on specification, estimation, and evaluation of trade elasticities. Their approach is primarily methodological, focusing on trade equations derived from alternative behavioral models. They consider recent advances in econometrics for evaluating the choice between competing models and provide examples of the use of some of these techniques in international trade.

In the second paper, Philip Abbott discusses econometric and economic issues concerning estimation of U.S. agricultural export demand elasticities. He points out the variety of problems researchers confront in trying to estimate export demand elasticities and describes solutions to some of these problems. He concludes with a discussion of two recent experiences in estimating excess demand elasticities for wheat and coarse grains.

The third paper, by John Dutton and Tom Grennes, analyzes the effect of exchange rate changes on the volume of trade. The authors begin with a discussion of factors common to prices and exchange rates as they effect trade, followed by a review of the literature on exchange rates. They then present estimation results for an aggregate U.S. agricultural export equation and for 3 separate crop exports--corn, wheat, and soybeans.

In the fourth paper, Irma Adelman and Sherman Robinson investigate the effects of macroeconomic shocks, foreign trade, and structural adjustments on the U.S. economy for the period 1982-86. Their tool of analysis is a multisectoral computable general equilibrium model of the United States. The model contains seven agricultural sectors and is designed to measure the impacts of exchange rate policy, trade embargoes, import quotas, tariffs, and production subsidies.

The fifth paper, by Christine Bolling, addresses the issues of price and exchange rate transmission in the case of Latin American countries. The paper begins with a comparison of producer price behavior in each Latin American country with respect to a representative world price for various commodities. Each country's exchange rate with respect to the U.S. dollar is also evaluated. Next, price transmission equations, both in nominal and real price terms, are derived based on a simple price linkage equation. The equations are then estimated and tested for statistical significance to determine how each country's farm prices responded to changes in the world price, the exchange rate, and the consumer price index.

In the sixth paper, David Blandford analyzes trade behavior of major wheat and corn importing countries. He applies Markov models for the importing countries over the period 1970 to 1983 using both annual and quarterly data. Price elasticities of market shares are derived for both the annual and quarterly models and tested for statistical difference.

The seventh paper, by Ellen Goddard, presents a theoretical framework for estimating import demand elasticities for differentiated products. This method is tested for the case of the market for beef products.

In the final paper, Rod Tyers and Kym Anderson derive trade elasticities from a multi-commodity world food model developed for analyzing the implications of trade liberalization. The model contains elasticities for international price transmission, domestic production, and domestic consumption for each country or region in the model.

Christina Fitz Gibbon and Laura Bipes were instrumental in assisting us in the completion of this book and we would like to thank them. Christina prepared the camera-ready copy and Laura handled the logistics of the Dearborn symposium where the papers in this volume were presented.

Colin A. Carter
Walter H. Gardiner

Chapter 1

Walter H. Gardiner and Colin A. Carter

Issues Associated with Elasticities in International Agricultural Trade

INTRODUCTION

One of the most important topics in international trade research is the responsiveness of trade flows to price changes, that is, price elasticity. This topic has been the subject of much theoretical and empirical investigation over the past 50 years because of the pivotal role it plays in the solution to a wide range of economic problems and policy issues.

The purpose of this paper is to provide an overview of issues associated with elasticities in international agricultural trade as background for a set of more detailed papers that are contained in this volume. The emphasis is on price elasticities since it is this parameter that has been the source of much debate in the agricultural economics profession over the years.

THE IMPORTANCE OF TRADE ELASTICITIES

The theoretical and practical importance of elasticities in international trade research is without issue. Estimating of elasticities has long been an endeavor of researchers of international trade (Cheng). Knowledge of these parameters provides a means for testing economic theories, explaining the structure of commodity markets, forecasting trade flows, and analyzing the effects of government policies. This information, when provided to policymakers, represents an important contribution to the process of designing and evaluating policy options.

International trade theory has provided a wealth of hypotheses to be tested and trade elasticities have played a crucial role in testing these hypotheses. Elasticities have been used to demonstrate the possibility that low price response for imports and exports can cause a currency devaluation to lead to a worsening of a country's trade imbalance. This led to the development of the Marshall-Lerner condition which states that a country's trade balance can be improved by depreciating its currency if the sum of the elasticities of the country's demand for imports and demand by the rest of the world for its exports is greater than one in absolute value (Grubel). Many of the attempts to estimate price elasticities in international trade during the 1950s and 1960s were designed to test the existence of the Marshall-Lerner condition or to evaluate other balance of payments issues (Leamer and Stern).

Elasticities have also been used to test theories relating to the commodity composition of trade (Valentini and Schuh), spatial equilibrium (Takayama and Judge), the "law of one price" (Grennes, Johnson, and Thursby), market structure (Carter and Schmitz), and game theory (Paarlberg).

One of the primary reasons for estimating elasticities in agricultural trade has been to increase the understanding of the structure and performance of commodity markets. During the 1950s and 1960s, researchers became increasingly interested in markets for primary commodities. This interest in commodity modeling was attributed to a sharp increase in the demand for primary commodities stemming from the economic growth in industrial countries after World War II. Commodity models provided the basis for estimating investment, production, and buying and selling decisions of market participants (Labys, 1973; Adams and Behrman, 1978).

Elasticities have also been used to construct agricultural models for forecasting and policy analysis. Forecasting models have been designed to predict future values of exports, imports, market shares, and prices (Chen; Scherr). Policy applications have included the impacts of both tariff and nontariff measures. Elasticities of import demand have long been used to estimate the effect of tariffs on trade and prices (Schultz). During the 1930s, the Wisconsin Tariff Commission with the assistance of the Rawleigh Foundation prepared tariff studies on sugar, feed grains, pork, mutton, dairy products, and wool (Renne). More recently, a number of researchers have synthesized price elasticities from existing studies to develop simulation models of agricultural markets (Ray and Richardson; Sharples, 1979; Holland and Meekhof; Miller and Washburn; Quance; Chattin, Hillberg, and Holland). These models have been used to address a wide range of domestic and foreign policy issues.

The world food crisis in the early 1970s and the subsequent expansion of the export market sparked a renewed interest in empirical research in international agricultural trade (Thompson; Rausser). A variety of methods for modeling and forecasting agricultural trade were developed. Despite numerous attempts to directly estimate price elasticities for agricultural exports, these studies failed to reach a consensus on the magnitude of the elasticity estimates (Thompson; Schmitz, McCalla, Mitchell, and Carter; Gardiner and Dixit).

Events in the early 1980s caused agricultural markets, particularly in the United States, to go from boom to bust. U.S. farm programs, designed to raise farm incomes primarily through price supports, were formulated in an earlier period when commercial trade was much less important for U.S. agriculture than it is today. U.S. farm policy now affects the international market in a major way.

U.S. loan rates for agricultural commodities rose steadily from 1976 through 1983. Rising loan rates acted as a floor for export prices and together with a strengthening U.S. dollar raised world market prices. Higher world prices enabled competing exporters, such as Australia, Argentina, Canada, France, South Africa, and Thailand, to increase production and provide more competition for the United States. Higher world prices also reduced import demand and encouraged production in importing countries which further cut into export sales. As a result, U.S. agricultural exports declined from a peak of $43.78 billion in 1980/81 to $26.32 billion in 1985/86 (USDA, Ag. Outlook, July 1987, p. 46). Some observers believed that if U.S. prices were lowered then exports would rebound to their levels in the latter part of the 1970s.

During the debate over the 1985 Food Security Act, short-term export price elasticities in excess of -1.0 were used to justify reduced loan rates and the marketing loan program as a means of regaining export market shares (FAPRI #3-86, Dec. 1986, p.10). Based on an assessment of the export markets for U.S. agricultural products one year after the 1985 Food Security Act took affect, the Food and Agricultural Policy Research Institute (FAPRI) suggested that short term elasticities, when corn, wheat, and soybean prices are declining together, are in the -0.3 to -0.5 range and the longer-term elasticities are in the range of -1.0 to -1.5 (Ibid.).

Thus, the successfulness of U.S. farm policy depends critically on the elasticity of demand for U.S. exports. If export demand is elastic, then the programs to increase exports through a lowering of world prices will be successful. On the other hand, if the international demand is inelastic, then export growth will be slow and the programs will be costly.

Whether or not the foreign demand for U.S. agricultural commodities is price responsive is a fundamental economic question. But it is a question to which there is no clear answer despite the substantial amount of research on the topic. It is fair to say there is a lot of confusion in the literature, not only on the empirical estimates of elasticities but also on the concept and the various definitions of it. The topic of elasticities in international trade is also of interest to many other trading nations besides the United States, especially the group of 14 exporting nations known as the CAIRNS group which has been protesting the EC and U.S. export subsidy programs. Furthermore, agricultural trade liberalization is being taken seriously for the first time in the Uruguay round of GATT negotiations, and member countries are interested in the effects of price changes on agricultural trade.

SPECIFICATION ISSUES

The empirical estimation of price elasticities in international trade begins with the specification of a model. Model specification raises a number of methodological issues including selection of an appropriate model, choice of variables, and the functional form of the model. The appropriate model depends on a number of factors:

1. the model's purpose -- hypothesis testing, structural analysis, forecasting, or policy analysis;
2. the nature of the commodity under investigation -- Is the commodity homogeneous so there exist close or perfect substitutes or is the commodity sufficiently differentiated so there are no perfect substitutes?
3. the type of market that the commodity is traded in -- competitive or imperfectly competitive; and
4. the desired degree of regional disaggregation.

The two models that have dominated the empirical literature on trade are the perfect substitutes model and imperfect substitutes model (Goldstein and Kahn). The principal distinction between the two models is that imports and exports in the perfect substitutes model are simply the residual or difference between domestic supply and demand functions, while the imperfect substitutes model contains separate behavioral functions for imports and exports. For the perfect substitutes model, elasticities for imports and exports are obtained by combining the domestic supply and demand elasticities with information on the level of imports and exports and domestic production and consumption. For the imperfect substitutes model, the price elasticities corresponding to imports and exports are obtained directly from the estimated parameters corresponding to the price variables.

Examples of both models abound in the literature and are well documented in a number of surveys (Labys, 1973; Adams and Behrman, 1978; Schmitz; Schuh, 1979; Sarris, 1981; Thompson; Schmitz, McCalla, Mitchell, and Carter; Gardiner and Dixit). Some specific modeling issues addressed by researchers include: price equilibrium versus price disequilibrium (Baumes and Womack), residual supplier (Bredahl and Green), and imperfect markets (McCalla and Josling).

Once a model has been chosen to represent the trade in a particular commodity or product group, the choice of appropriate variables for the model becomes an issue. Economic theory tailored with information about the distinguishing features of the agriculture sector guides the researcher in this selection process. One of the main problems in specifying import and export functions for direct estimation is deciding which price variables to include to capture the direct price response and substitution relationships. This will depend in a large part on the level of aggregation -- both commodity and country. A country's import demand equation for total agriculture might include a price (index or unit value) for agricultural imports, a domestic price for agricultural goods, and prices for nonagricultural goods. Alternatively, if imports (or exports) are disaggregated by commodity or by country, there is an increase in the number of competitors and the number of price and other variables to consider. This can lead to a variety of estimation problems.

One solution to the problem is a methodology proposed by Armington which views the import decision process as occurring in two stages. This method permits the calculation of the cross-price elasticities between all exporters from estimates of the aggregate price elasticity for imports, the elasticity of substitution, and trade shares. Early attempts to apply the Armington framework to agricultural trade were made at North Carolina State University by Wells; K. Collins (1977); and Johnson, Grennes, and Thursby (1977). The popularity of this approach is reflected in the large number of Armington models that have been built since the late 1970s (Grennes, Johnson, and Thursby, 1978; Sarris, 1983; Honma and Heady; Alston; Abbott and Paarlberg; Dixit and Roningen; Figueroa; Figueroa and Webb; Suryana; Haniotis and Ames).

The simplistic and restrictive assumptions of the Armington model have been challenged by Winters. In particular, Winters found the Armington assumptions of separability between domestic and foreign goods and between all import sources, to be unacceptable. He found price response in the Armington model to be highly constrained. Instead, Winters uses a model called the "almost ideal demand system" (Deaton and Muellbauer) to test the Armington assumptions and to explain international trade flows in OECD manufacturers. The Armington assumptions of homotheticity and separability are rejected.

A study prepared for Agriculture Canada by Intercambio Ltd. modeled the world wheat market using the almost ideal demand system. Wheat was not treated as a homogeneous product, rather it was divided into different classes. The own price elasticity, the cross price elasticities, and the income elasticities were estimated and found to be relatively large.

An important characteristic of price elasticities is that they are temporal and therefore change with time. Imports and exports do not adjust instantly to price changes or other explanatory variables due to time delays or lags that result from imperfect information or other adjustment costs. Junz and Rhomberg identify five different time delays or lags between price changes and changes in quantities traded -- the recognition lag, the decision lag, the delivery lag, the replacement lag, and the production lag. The length of these time lags will vary from commodity to commodity, and will depend on the underlying factors which determine the response of producers and consumers to price changes -- the nature of the biological production process; the perishability of the commodity; technology in production, processing, and marketing; government policy; and the number and kind of available substitutes. Thus commodity trade behavior in many cases is more appropriately described by a dynamic response function rather than a static one.

The issue in specifying trade equations is not whether time lags ought to be considered but rather how best to account for the time period of adjustment. The specification of dynamic commodity relationships is rooted in separate early works by Koyck and Nerlove. Procedures for handling response lags in trade equations are discussed in Leamer and Stern, Labys (1973), and Goldstein and Kahn. Examples of agricultural trade equations which use some form of dynamic specification include studies by Witherell for wool; Labys (1975a) for lauric oils; Kost, Schwartz and Burris for coarse grains; and Chambers and Just (1981) and Conway for corn, wheat, and soybeans.

Another important specification issue in agricultural trade research is the treatment of exchange rates in trade equations. The importance of exchange rates and its effect on agriculture was highlighted in a classic paper by Schuh (1974) and has spawned over a decade of research on this subject. In their review of the literature on exchange rates and agricultural markets, Chambers and Just (1979) argue that the usual approaches of handling exchange rates in agricultural trade models -- excluding them or using them to adjust prices --were too restrictive, leading to biased parameter estimates. They suggest a more general specification which include a separate exchange rate variable in the equation to be estimated. Their paper, like Schuh's, generated a great deal of controversy and additional research in the profession. The treatment of exchange rates in agricultural trade was a major topic of discussion at the 1986 summer symposium of the International Agricultural Trade Research Consortium (Chambers and Paarlberg).

ESTIMATION ISSUES

Estimates of trade elasticities from the 1930s and 1940s were often quite low (inelastic) in value. This along with the fact that some countries experienced deterioration in their trade balances following a currency devaluation led to the contention of "elasticity pessimism" among economists and government officials (Grubel). Studies in the early 1950s by Machlup; Orcutt; and Harberger pointed out a variety of problems associated with estimation of price elasticities in international trade which could lead to an underestimation of the true price response. The list of problems include: specification error, simultaneous equation bias, identification problems, measurement errors, aggregation problems, inappropriate lag structure, and scale effects.

The presence of a number of these problems in agricultural trade research are acknowledged by Schmitz; Sarris (1981), and Thompson in their surveys of agricultural trade models, particularly as they related to the large number of attempts in the 1970s to directly estimate agricultural export demand equations. The problem of simultaneity between prices and the error term can be quite damaging to estimates of excess demand functions for agricultural commodities because large shifts in agricultural production due to environmental factors imply large shifts in the excess demand functions. Thus, the use of single-equation estimation procedures such as ordinary least squares (OLS) will yield biased and inconsistent parameter estimates. The recognition of the simultaneity problem brought forth calls for use of simultaneous equation estimators. Binkley and McKinzie address the problem of simultaneous equation bias in more detail with a Monte Carlo analysis in which they compare the relative merits of alternative methods for estimating export demand functions. They provide the researcher with some guidelines for choosing among various estimation methods when estimating trade equations.

One of the earliest attempts to estimate elasticities that was designed to be free of simultaneous equation bias was a study by Morgan and Corlett. The authors use limited information maximum likelihood methods to estimate import demand equations for various commodities over the period 1890-1914. While there were other attempts to use simultaneous equations methods in the 1960s and the 1970s, most estimates of agricultural import and export equations were made with single equation methods like OLS. However, since the late 1970s, there appears to have been greater use made of simultaneous equation estimators in international commodity modeling.

A number of aggregation issues arise in the process of estimating elasticities in international trade. In the case of price elasticities, is it necessary to obtain estimates at the disaggregated level (i.e., individual commodity and/or country) and combine them to get an aggregate price elasticity or can the aggregate elasticity be obtained directly?

This issue was addressed by Orcutt and later analyzed in greater detail by Theil. In the late 1970s, a lively debate emerged over the aggregation problems associated with calculating aggregate elasticities of demand for agricultural exports. Earlier, in his article on demand for U.S. farm output, Tweeten (1967) calculated the price elasticity of export demand for U.S. food and feed with a value of -6.4 by aggregating across individual countries' and regions' demand and supply elasticities. In an article 10 years later, Johnson (1977) criticized Tweeten's method for aggregating across countries and suggested the correct procedure could be found in Floyd (1965) or Johnson (1970). Johnson chose instead to aggregate across commodities and obtained an elasticity for U.S. farm exports of -6.69 which is in substantial agreement with Tweeten's highly elastic estimate. Tweeten (1977) responded to Johnson's criticism by deriving the equation for the elasticity of export demand and explained his calculation procedure. Tweeten concluded that his procedure was not in error and was conceptually superior to Johnson's procedure.

Another issue that arose out of this debate was the effect of both domestic and trade policies on elasticity estimates. Bredahl, Meyers and Collins claim that Tweeten and Johnson did not consider foreign government policies which insulate producers and consumers from international price movements. This is implied by their assumption of perfect price transmission. Bredahl, Meyers and Collins computed export demand elasticities for various U.S. agricultural commodities under three alternative assumptions regarding the price transmission elasticities for various countries. They concluded that the lower value of their elasticity estimates explained the discrepancy between Tweeten and Johnson's highly elastic values and most other empirical estimates which were inelastic or slightly greater than one. Price transmission elasticities have since been widely used in empirical work on agricultural trade (Collins, 1980; Dunmore and Longmire; Seeley; McCalla, Abbott, and Paarlberg; Roe, Shane, and Vo; Tyers and Anderson; USDA, 1986; Magiera; Roningen, Sullivan, and Wainio).

Abbott (1979) also critiqued the existing trade estimation methodologies, noting that deviations of domestic prices from international prices were either ignored or entered exogenously in the model. He suggested a modified approach to modeling international trade flows that would improve estimates of price response by incorporating government policies directly in the estimation process. Roe, Shane, and Vo recently developed a formal model of endogenous government behavior to analyze the effect of government intervention on import behavior. They specified a government choice function and derived the implied import demand functions which they estimated using cross-country and time-series data. Using an estimated import equation, the authors measured the responsiveness of import demand and excess demand to price changes and the extent to which price changes are transmitted to final consumers.

REFERENCES

Abbott, P. C. "Modeling International Grain Trade with Government Controlled Markets," American Journal of Agricultural Economics 61(1979):22-31.

_________. "Estimating U.S. Agricultural Export Demand Elasticities: Econometric and Economic Issues," paper presented at the International Agricultural Trade Research Consortium Analytical Symposium, Dearborn, Michigan, July 30-August 1, 1987.

Abbott, P. and P. Paarlberg. "Modeling the Impact of the 1980 Grain Embargo." Embargoes, Surplus Disposal, and U.S. Agriculture, Staff Report No. AGES860910, Chapter 11, U.S. Department of Agriculture, Economic Research Service, November 1986.

Adams, F. G. and J. R. Behrman. Econometric Modeling of World Commodity Policy, Lexington, Mass: Lexington Books, D.C. Heath, 1978.

________________________. Econometric Models of World Agricultural Commodity Markets, Cambridge, Mass: Ballinger, 1976.

Adelman, I. and S. Robinson. "Macroeconomic Shocks, Foreign Trade, and Structural Adjustments: A General Equilibrium Analysis of the U.S. Economy, 1982-86," paper presented at the International Agricultural Trade Research Consortium Analytical Symposium, Dearborn, Michigan, July 30-August 1, 1987.

Alston, J. M. "The Effects of the European Community's Common Agricultural Policy in International Markets form Poultry Meat," North Carolina Agr. Res. Bul. 471, North Carolina State University, March 1985.

Armington, P. S. "A Theory of Demand for Products Distinguished by Place of Production," IMF Staff Papers 16(1969):159-178.

Arnade, C. A. and C. W. Davison. "Export Demand for U.S. Wheat," paper presented at American Agricultural Economics Association meeting, Reno, Nevada, July 29, 1986.

Baumes, H. S. and W. H. Meyers. "The Crops Model: Structural Equations, Definitions and Selected Impact Multipliers," NED Staff Paper. U.S. Department of Agriculture, Economics, Statistics, and Cooperative Services, March 1980.

Baumes, H. S. and A. W. Womack. "Price Disequilibrium versus Price Equilibrium: A Question of Modeling Approaches," Unpublished working paper. U.S. Department of Agriculture, Economic Research Service, July 1979.

Binkley, J. K. and L. D. McKinzie. "A Monte Carlo Analysis of the Estimation of Export Demand," Canadian Journal of Agricultural Economics 29(1981):187-202.

Blandford, D. "Market Share Models and the Elasticity of Demand for U.S. Agricultural Exports," paper presented at the International Agricultural Trade Research Consortium Analytical Symposium, Dearborn, Michigan, July 30-August 1, 1987.

Bolling, C. "Price and Exchange Rate Transmission Revisited: The Latin American Case," paper presented at the International Agricultural Trade Research Consortium Analytical Symposium, Dearborn, Michigan, July 30-August 1, 1987.

Bredahl, M. E., P. Gallagher, and J. Matthews. "Aggregate Export Demand for U.S. Agricultural Commodities: Theory and Implications for Empirical Research," Forecast Support Group Working Paper, Commodity Economics Division, Economics, Statistics, and Cooperative Services, U.S. Department of Agriculture, February 1978.

Bredahl, M. E., W. Meyers, and K. J. Collins. "The Elasticity of Foreign Demand for U.S. Agriculture Products: The Importance of the Price Transmission Elasticity," American Journal of Agricultural Economics 61(1979):58-62.

Bredahl, M. E. and L. Green. "Residual Supplier Model of Coarse Grains Trade," American Journal of Agricultural Economics 65(1983):785-790.

Capel, R. E. and L. R. Rigaux. "Analysis of Export Demand for Canadian Wheat," Canadian Journal of Agricultural Economics 22(1974):1-15.

Carter, C. and A. Schmitz. "Import Tariffs and Price Formation in the World Wheat Market," American Journal of Agricultural Economics 61(1979):517-522.

Chambers, R. G. and R. E. Just. "A Critique of Exchange Rate Treatment in Agricultural Trade Models," American Journal of Agricultural Economics, 61(1979):249-257.

Chambers, R. G. and R. E. Just. "Effects of Exchange Rate Changes on U.S. Agriculture: A Dynamic Analysis," American Journal of Agricultural Economics 63(1981):32-46.

Chambers, R. G. and P. L. Paarlberg (editors), Chattin, B. L., A. M. Hillberg, and F. D. Holland. PC WHEATSIM: Model Description and Computer Program Documentation, Purdue University, Station Bul. No. 477, September 1985.

Chen, D. T. "The Wharton Agricultural Model: Structure, Specifications, and Some Simulation Results," American Journal of Agricultural Economics 59(1977):107-16.

Cheng, H. S. "Statistical Estimates of Elasticities and Propensities in International Trade - A Survey of Published Studies," IMF Staff Papers 7(1959):107-158.

Collins, H. Christine. "Price Transmission Between the U.S. Gulf Ports and Foreign Farm Markets," IED Staff Report, Economics, Statistics, and Cooperative Services, U.S. Department of Agriculture, January 1980.

________________. "Price and Exchange Rate Transmission," Agricultural Economics Research 32(1980):50-55.

Collins, K. J. "An Economic Analysis of Export Competition in the World Coarse Grains Market: A Short-Run Constant Elasticity of Substitution Approach," Ph.D. dissertation. North Carolina State University, Raleigh, NC, 1977.

Conway, Roger K. "Examining Intertemporal Export Elasticities for Wheat, Corn, and Soybeans: A Stochastic Coefficients Approach," Tech. Bul. 1709, U.S. Department of Agriculture, Economic Research Service, October 1985.

Cronin, M. R. "Export Demand Elasticities with Less Than Perfect Markets," Australian Journal of Agricultural Economics 23(1979):69-72.

Deaton, A. S. and J. Muellbauer. "An Almost Ideal Demand System," American Economic Review 70(1980):312-26.

Dixit, P. M. and V. O. Roningen. "Modeling Bilateral Trade Flows with the Static World Policy Simulation (SWOPSIM) Modeling Framework," Staff Report No. AGES861124, U.S. Department of Agriculture, Economic Research Service, December 1986.

Dunmore, J. C. and J. Longmire. "Sources of Recent Changes in U.S. Agricultural Exports," ERS Staff Report No. AGES831219. U.S. Department of Agriculture, Economic Research Service, January 1984.

Dutton, J. and T. Grennes. "The Role of Exchange Rates in Trade Models," paper presented at the International Agricultural Trade Research Consortium Analytical Symposium, Dearborn, Michigan, July 30-August 1, 1987.

Figueroa, E. "Implications of Changes in the U.S. Exchange Rate for Commodity Trade Patterns and Composition," Ph.D. dissertation, University of California, Davis, 1986.

Figueroa, E. and A. J. Webb. "An Analysis of the U.S. Grain Embargo Using a Quarterly Armington-Type Model." Embargoes, Surplus Disposal, and U.S. Agriculture, Staff Report No. AGES860910, Chapter 12, U.S. Department Agriculture, Economic Research Service, November 1986.

Floyd, J. B. "The Overvaluation of the Dollar: A Note on the International Price Mechanism," American Economic Review 55(1965):95-107.

Food and Agricultural Policy Research Institute, "The Food Security Act of 1985 One Year Later: Implications and Persistent Problems," FAPRI Staff Report #3-86, University of Missouri-Columbia and Iowa State University, December 1986.

Gardiner, W. H. and P. M. Dixit. "Price Elasticity of Export Demand: Concepts and Estimates," FAER No. 228, U.S. Department Agriculture, Economic Research Service, February 1987.

Goddard, E. "Export Demand Elasticities in the World Market for Beef," paper presented at the International Agricultural Trade Research Consortium Analytical Symposium, Dearborn, Michigan, July 30-August 1, 1987.

Goldstein, Morris and Moshin S. Kahn. "Income and Price Effects in Foreign Trade," Handbook of International Economics, Vol. II, edited by R.W. Jones and P. B. Kenen, Elsevier Science Publishers B.V., 1985.

Grennes, T., P. R. Johnson, and M. Thursby. The Economics of World Grain Trade, New York: Praeger Publishers, 1977.

Grubel, H. G. International Economics, Homewood, IL: Richard D. Irwin, 1977.

Haniotis, T. and Glen C. W. Ames. "European Community Enlargement: Impact on U.S. Soybean Exports to the EC," paper presented at the Southern Agricultural Economics Conference, Nashville, TN, February 1987.

Harberger, A. C. "A Structural Approach to Problems of Import Demand," American Economic Review 43(1953):148-159.

Holland, F. D. and R. L. Meekhof. FEEDSIM: Description and Computer Program Documentation, Station Bulletin No. 221, Purdue University, March 1979.

Holland, F. D. and J. A. Sharples. WHEATSIM: Model 15 Description and Computer Program Documentation, Station Bulletin No. 319. Purdue University, March 1981.

Honma, M. and E. O. Heady. An Econometric Model for International Wheat Trade: Exports, Imports, and Trade Flows, CARD Report 124, Center for Agricultural and Rural Development, Ames, IA, 1984.

Houthakker, H. S. and S. P. Magee. "Income and Price Elasticities in World Trade," Review of Economics and Statistics 51(1969):111-125.

Intercambio Ltd. "Analysis of Strategic Mixes for Canadian Wheat Exports," studied prepared for Agriculture Canada, Ottawa, Canada, 1985.

Johnson, P. R. "Balance of Payments 'Pressure: The Columbian Case," Southern Economic Journal 37(1970):163-73.

___________. "The Elasticity of Foreign Demand for U.S. Agricultural Products," American Journal of Agricultural Economics 59(1977):735-36.

Johnson, P. R., T. Grennes, and M. Thursby. "Trade Models with Differentiated Products," American Journal of Agricultural Economics 61(1979):120-127.

_______________________________. "Devaluation,Foreign Trade Controls, and Domestic Wheat Prices," American Journal of Agricultural Economics 59(1977):619-27.

Junz, H. B. and R. R. Rhomberg. "Price Competitiveness in Export Trade Among Industrial Countries," American Economic Review, 63(1973):412-18.

Konandreas, P., P. Bushnell, and R. Green. "Estimations of Export Demand for U.S. Wheat Exports." Western Journal of Agricultural Economics 3(1978):39-49.

Kost, W. E., M. Schwartz, and A. Burris. "FDCD World Trade Forecast Modeling System," unpublished working paper. U.S. Department of Agriculture, Economic Research Service, August 1979.

Koyck, L. M. Distributed Lags and Investment Analysis, North-Holland Publishing Co., Amsterdam, 1954.

Labys, W. C. Dynamic Commodity Models: Specification, Estimation, and Simulation, Lexington Books, Heath & Co., Lexington, Mass., 1973.

_________. "Dynamics of the Lauric Oil Market" in W. C. Labys (ed.), Quantitative Models of Commodity Markets, Cambridge: Ballinger Pub. Co., 1975a.

_________. "The Problems and Challenges for International Commodity Models and Model Builders," American Journal of Agricultural Economics 57 (1975b):873-878.

Labys, W. C. and P. K. Pollak. Commodity Models for Forecasting and Policy Analysis, Croom Helm Ltd, London, 1984.

Leamer, E. E. and R. M. Stern. Quantitative International Economics, Allyn and Bacon, Boston, 1970.

Machlup, F. "Elasticity Pessimism in International Trade," Economia Internazionale 3(1950):118-137.

Magee, S. P. "Prices, Income and Foreign Trade: A Survey of Recent Economic Studies," in: P. B. Kenen, ed., International Trade and Finance: Frontiers for Research, Cambridge University Press, Cambridge, 1975.

Magiera, Stephen L. "Trade Conflict or Negotiations: Scenarios for Developed Country Adjustment to Structural Market Surpluses," forthcoming USDA report, 1987.

McCalla, A. and T. Josling, eds. Imperfect Markets in Agricultural Trade, Montclair, N.J.: Allanheld, Osmun and Co., 1981.

McCalla, A., P. Abbott, and P. Paarlberg. "Policy Interdependence, Country Response, and the Analytical Challenge," Embargoes, Surplus Disposal, and U.S. Agriculture, Staff Report No. AGES860910, Chapter 6, U.S. Department Agriculture, Economic Research Service, November 1986.

Miller, T. A. and M. C. Washburn. "AGSEM: An Agricultural Sector Equilibrium Model," CED Working Paper. U.S. Department Agriculture, Economics, Statistics, and Cooperative Services, April 1978.

Nerlove, M. Distributed Lags and Demand Analysis, Agricultural Handbook No. 141, U.S. Department of Agriculture, Economic Research Service, Washington, D.C., 1958.

Orcutt, G. H. "Measurement of Price Elasticities in International Trade," Review of Economics and Statistics 32(1950):117-132.

Orden, David. "A Critique of Exchange Rate Treatment in Agricultural Trade Models: Comment," American Journal of Agricultural Economics 68(1986):990-93.

Paarlberg, P. L. "Endogenous Policy Formation in the Imperfect World Wheat Market," Ph.D. dissertation. Purdue University, 1983.

Quance, C. L. "National Interregional Agricultural Projections (NIRAP) System: An Executive Briefing," working paper, U.S. Department of Agriculture, Economics, Statistics, and Cooperative Services, April 1980.

Rausser, G. C. "New Conceptual Developments and Measurements for Modeling the U.S. Agricultural Sector," New Directions in Econometric Modeling and Forecasting in U.S. Agriculture, ed. G. C. Rausser, p. 1-14. North Holland, 1982.

Ray, D. E. and J. W. Richardson. Detailed Description of POLYSIM, Tech. Bul. T-151. Agricultural Experiment Station, Oklahoma State University, December 1978.

Renne, R. R. "Verification of Tariff Effectiveness By Different Statistical Methods," Journal of Farm Economics 16(1934):591-601.

Roe, T., M. Shane, and D. H. Vo. "Price Responsiveness of World Grain Markets," Tech. Bul. No. 1720, Washington, DC: U.S. Department Agriculture, Economic Research Service, June 1986.

Roningen, V. O., J. Sullivan, and J. Wainio. "The Liberalization of Agricultural Support in the United States, Canada, the European Community, and Japan," paper presented at the GATT Agricultural Policy Modeling Workshop, London, Ontario, May 1987.

Ryan, T. J. "International Trade Models: An Overview," working paper, U.S. Department of Agriculture, Economics, Statistics, and Cooperative Services, March 1979.

Sarris, A. H. "Empirical Models of International Trade in Agricultural Commodities," in A. McCalla and T. Josling, eds., Imperfect Markets in Agricultural Trade, Montclair, N.J.: Allanheld, Osmun and Co., 1981, p. 87-112.

_________. "European Community Enlargement and World Trade in Fruits and Vegetables," American Journal of Agricultural Economics 65(1983):235-46.

Scherr, B. A. "The 1978 Version of the DRI Agriculture Model," Working Paper No. 5, Data Resources, Inc., Lexington, Mass., January 1978.

Schmitz, A. "Research in International Trade: Methods and Techniques--With Emphasis an Agricultural Trade," in J. Hillman and A. Schmitz, eds., Trade and Agriculture--Theory and Policy, Boulder, Co.: Westview Press, 1979, p. 273-294.

Schmitz, A., A. F. McCalla, D. O. Mitchell, and C. A. Carter. Grain Export Cartels. Cambridge, MA: Ballinger Publishing Co., 1981.

Schuh, G. E. "The Exchange Rate and U.S. Agriculture," American Journal of Agricultural Economics 56(1974):1-13.

_________. "Problems Involved in Doing Research on International Trade in the Agricultural Sector," in J. Hillman and A. Schmitz, eds., Trade and Agriculture--Theory and Policy, Boulder, Co.: Westview Press, 1979, p. 255-273.

_________. "The Foreign Trade Linkages," Modeling Agriculture for Policy Analysis in the 1980s. Federal Reserve Bank of Kansas City Symposium, Kansas City, MO, September 1981

Schultz, Henry. "Correct and Incorrect Methods of Determining the Effectiveness of the Tariff," Journal of Farm Economics 17(1935):625-41.

Seeley, R. "Price Elasticities from the IIASA World Agriculture Model," ERS Staff Report No. AGES850418. U.S. Department Agriculture, Economic Research Service, 1985.

Sharples, J. A. "WHEATSIM: Model for Analysis of Wheat Policy Alternatives," working paper, U.S. Department of Agriculture, Economics, Statistics and Cooperative Services, January 1979.

_________. "The Short-Run Elasticity of Demand for U.S. Wheat Exports," ERS Staff Report AGES820406. U.S. Department Agriculture, Economic Research Service, 1982.

Stern, R. M., J. Francis, and B. Schumacher. Price Elasticities in International Trade - An Annotated Bibliography, The Trade Policy Research Centre, Macmillan, London, 1976.

Suryana, A. "Trade Prospects of Indonesian Palm Oil in the International Markets for Fats and Oils," Ph.D. dissertation, North Carolina State University, 1986.

Takayama, T. and G. G. Judge. Spatial and Temporal Price and Allocation Models, Amsterdam: North Holland, 1971.

Theil, H. Linear Aggregation of Economic Relations, North-Holland, Amsterdam, 1954.

Thompson, R. L. "A Survey of Recent U.S. Developments in International Agricultural Trade Models," BLA-21, U.S. Department Agriculture, Economic Research Service, September 1981.

Thursby, J. G. and M.C. Thursby. "How Reliable are Simple, Single Equation Specifications of Import Demand", Review of Economics and Statistics 66(February 1984):120-128.

Thursby, J. G. and M.C. Thursby. "Elasticities in International Trade: Theoretical and Methodological Issues," paper presented at the International Agricultural Trade Research Consortium Analytical Symposium, Dearborn, Michigan, July 30-August 1, 1987.

Tryfos, P. "The Measurement of Price Elasticities in International Trade," American Journal of Agricultural Economics 57(1975):689-691.

Tweeten, Luther. "The Demand for United States Farm Output," Food Research Institute Studies 7(1967):343-69.

______________. "The Elasticity of Foreign Demand for U.S. Agricultural Products: Comment," American Journal of Agricultural Economics, 59(1977):737-38.

Tweeten, L. and S. Rastegari. "The Elasticity of Export Demand for U.S. Wheat." Mimeograph, Oklahoma State University, Department Agricultural Economics, 1984.

Tyers, R. and K. Anderson. "Distortions in World Food Markets: A Quantitative Assessment," unpublished manuscript, January 1986.

____________________. "Imperfect Price Transmission and Implied Trade Elasticities in a Multicommodity World Food Model," Chapter 9 in this volume.

U.S. Department of Agriculture. "Agricultural Trade Responsiveness in Western Hemisphere Countries," ERS Staff Report No. AGES860326, Economic Research Service, April 1986.

________________________. "Embargoes, Surplus Disposal, and U.S. Agriculture," Staff Report No. AGES860910, Economic Research Service, November 1986.

________________________. Agricultural Outlook, AO-132, Economic Research Service, July 1987.

Valentini, R. and G. E. Schuh. "The Metaproduction Function, Technology, and Trade in Agricultural Products," paper presented at the Annual Meeting of the Econometrics Society, San Francisco, December 1974.

Wells, Gary. "The Impact on Wheat Import Patterns from the Entrance of the United Kingdom into the European Economic Community," Ph.D. dissertation, Department of Economics and Business, North Carolina State University, 1977.

Winters, L. A. "Separability and the Specification of Foreign Trade Functions," Journal of International Economics (1984): 239-263.

Witherell, W. H. "Dynamics of the International Wool Market: An Econometric Analysis," Research Memo. No. 91, Econometric Research Program, Princeton University, 1967.

Chapter 2

Jerry G. Thursby and Marie C. Thursby

Elasticities in International Trade: Theoretical and Methodological Issues[1]

INTRODUCTION

Economists have devoted considerable attention to the estimation of international trade flows. Not only is there an extensive literature on the specification and estimation of equations describing these flows, but also there is a large literature surveying these studies (see, for example, Cheng (1959), Prais (1962), Kreinin (1967), Taplin (1967), Leamer and Stern (1970), Magee (1975), Stern et.al. (1976), Thompson (1981), Woodland (1982), Goldstein and Khan (1985), and Gardiner and Dixit (1986)).[2]

This is hardly surprising for several reasons. First there are wide-ranging positive, as well as normative, uses for estimated trade equations. Positive uses range from testing trade theories to understanding the transmission of economic disturbances across countries, and normative uses include evaluating alternative commercial policies, exchange-rate regimes, and macroeconomic policies.[3]

Second, time series data for international transactions have been easy to obtain historically, making empirical trade studies feasible. Somewhat paradoxically, the data rarely have been appropriate for estimating theoretically derived relationships, so that much of the existing work has focused on methods for dealing with errors in variables or omitted variables. Finally, there is wide variation in the estimated parameter values of trade equations. Price and income elasticities of demand and supply for imports and exports vary by commodity, country, and time period. Even when comparisons of estimates are limited to studies of narrowly defined commodities exported by a single country, estimates vary dramatically (see, for example, Gardiner and Dixit (1986)). Hence a major focus of existing surveys has been to catalogue and, to the extent possible, present a consensus of estimated elasticities.

This paper is a survey of recent research on specification, estimation and evaluation of trade elasticities. Since our focus is primarily methodological we do not give a compendium of recent estimates. Given the excellent and comprehensive nature of previous surveys, the marginal benefit of doing so would be small. In addition, we shall argue that any hope of obtaining a consensus of parameter values from trade equations must rely on taking a different approach. The approach involves using (and allowing the reader to use) as much information as is practically possible. There are both theoretical and econometric reasons to pursue such an approach, and we shall focus on studies which clarify them.

Section 2 outlines issues related to the theoretical specification of trade models or equations. We focus on differences in modelling suggested by differences in commodity substitutability and by whether they are purchased for final consumption or as inputs into a production process. This allows us to emphasize the fact that very different behavioral models can lead to the same estimating equation for trade flows, in which case proper interpretation of parameter estimates calls for estimation of a system of equations. Unfortunately, the bulk of the empirical trade literature reports single estimating equations with only cursory reference to the theoretical structure motivating the equations. Hence, even if estimated parameters were "reliable," it would be difficult to judge the usefulness of the estimates.

Section 3 considers recent advances in econometric techniques appropriate for assessment of a given model as regards its specification and to the choice of competing models. Section 4 gives several examples of the use of several of these techniques in international trade. One example comes from a study of trade aggregated over commodities, and the other considers Japanese demand for wheat imports from the United States. Finally, Section 5 concludes.

THEORETICAL SPECIFICATION OF TRADE EQUATIONS

In this section we focus on issues related to the theoretical specification of trade equations, where the term trade equation means the demand or supply equation for either exports or imports. While examples of each of these can be found in the literature, by far the major emphasis has been on demand equations.[4]

The economics literature has focused primarily on import demand functions, while the agricultural economics literature has focused on export demand functions (see, for example, any of the listed surveys of the economics literature and Gardiner and Dixit (1986) regarding the agricultural economics literature).

In order to discuss the different trade equations together, it is useful to think of them in the following context. For any commodity, a country's net trade can be represented by:

$$e_i = S_i\left(p_i^s, f_i\right) - D_i\left(p_i^d, y_i\right) \qquad (1)$$

where i is a country index, i = 1,...,m, p_i^s is a vector of supply prices, f_i is a vector of factor costs (or factor endowments), p_i^d is a vector of consumer prices, and y_i is income. $S_i(\bullet) \gtreqless 0$ denotes net supply by domestic producers and can be derived theoretically from either a technology or cost function (See Woodland (1982) and references therein). $D_i(\bullet)$ denotes domestic consumer demand and can be derived theoretically by constrained utility maximization.

Equation (1) can be used to represent excess supply for either a single commodity or aggregate commodity trade. The economics literature has dealt largely with estimates of aggregate trade, while the agricultural economics literature has focused on more narrowly defined commodities. We shall abstract from whether the commodity of interest is an aggregate, except to note that there is a literature dealing with when aggregation is appropriate (Green (1964), Berndt and Christensen (1973)).

If e_i is positive, country i is a net exporter of the commodity, and equation (1) can be used to describe its export supply. Country i is a net importer of commodities for which e_i is negative, and $(-e_i)$ can be used to represent the demand for imports of these commodities. $\sum_i e_i = 0$ in equilibrium for any commodity. Hence, in equilibrium, if country i is a net exporter of a commodity, its export demand can be represented by $\sum_{j \neq i} (-e_j)$. If country i is a net importer, it faces export supply given by $\sum_{j \neq i} e_j$.

Perfect Substitutes

This is the usual representation of trade equations when imports and exports are perfect substitutes for goods produced and consumed domestically. It has the convenient property of allowing trade elasticities to be calculated from domestic demand and supply elasticities (Yntema (1932), Johnson (1977), Goldstein and Khan (1985)). For example, a country's price elasticity of demand for imports of a commodity can be expressed as

$$\eta^d_{im} = (S_i/e_i)(d\ln S_i/d\ln p_i) - (D_i/e_i)(d\ln D_i/d\ln p_i) \qquad (2)$$

This expression follows from differentiating $(-e_i)$ from (1), where domestic subsidies and taxes are assumed to be zero so that $p^s_i = p^d_i$. Similarly, the price elasticity of demand for country i's exports of a commodity can be expressed as

$$\eta^d_{ix} = \eta^d_{jm} = \left(d\ln p_j/d\ln p^e_i\right)\left[\left(D_j/e_i\right)\left(d\ln D_j/d\ln p_j\right) - \left(S_j/e_i\right)\left(d\ln S_j/d\ln p_j\right)\right] \qquad (3)$$

where j is an aggregate of all countries other than i (thus $e_i = -e_j$), p^e_i is the export price, and again, internal subsidies and taxes are zero. The first term on the right hand side of (3) is a transmission elasticity showing the impact of a change in i's export price on the price faced by importers in j. While it is typically assumed to be unity, nontariff barriers or specific tariffs will lower its value (Goldstein and Khan (1985), Gardiner and Dixit (1986)).

Similar expressions can be derived for income elasticities of demand and for supply elasticities for imports and exports when goods are perfect substitutes (see Magee (1975)). Hence a researcher may choose to use domestic elasticity estimates to calculate trade elasticities rather than estimate them directly. Examples of studies using this method in agricultural economics are given in Gardiner and Dixit (1986). Its use is limited in obtaining elasticities for manufactures or aggregate trade since comparable estimates of domestic elasticities are rarely available, and these commodity classifications are not considered to be homogeneous.[5]

Outside the agricultural economics literature, empirical studies of trade have tended to adopt an imperfect substitutes framework. An exception to this is Clements' (1980) multisector econometric model of a small, open economy. His model explains production, consumption, and trade of three goods: exportables, importables, and non-traded goods. The net trade equations are similar to (1) with the net supply and demand functions being derived from optimizing behavior of producers and consumers. The model is applied to the United States for the period 1952-71. Implementing such a model is an ambitious task, but the benefit of the approach is that trade elasticities are estimated in a context where their theoretical interpretation is clear. As we discuss below, recent work in imperfect substitutes models has emphasized the need for such an approach.

Imperfect Substitutes

When internationally traded goods are not close substitutes for domestically traded goods, it is conventional to drop the representation of trade equations as the excess between domestic supply and demand.[6] In this case, demand for imports (exports) is typically written as a function of a price vector and income of the appropriate country (regional aggregate).[7] The simplest examples are

$$D_i^m = f\left(p_i^m, p_i^o, y_i\right) \tag{4}$$

$$D_i^x = D_j^m = g\left(p_j^m, p_j^o, y_j\right) \tag{5}$$

where, again, j is an aggregate of all countries but i, D_i^m and D_i^x denote import and export demand, p_i^m and p_j^m denote the price of imports in i and j (rest of the world), p_i^o and p_j^o denote price indexes for domestically produced goods in i and j, and y denotes real national income. More often than not, supply is assumed to be infinitely elastic; but several studies have specified export (import) supply as a function of an appropriate price vector and activity or capacity variable.[8]

In the majority of studies, specifications are chosen according to issues related to estimation, with little attention paid to the behavioral models underlying them. The demands specified in (4) and (5) are often presumed to come from utility maximization, but authors rarely state how the exact functions estimated are derived. As a result, functions may actually be inconsistent with the presumed theory. For example, the log-linear form of (4) and (5) is popular because the parameter estimates can be interpreted as elasticities, but it is not derivable from constrained utility maximization. Efforts to correct this deficiency have been made by Gregory (1971), Burgess (1974a, 1974b), and Kohli (1978, 1982). Gregory (1971) derives an equation for the ratio of imports to domestic goods as a function of their relative prices under the assumption that society's preferences can be represented by a CES utility function. The work of Burgess and Kohli has focused on two issues: (i)-the appropriateness of separability restrictions implied by typically estimated trade equations,[9] and (ii)-the derivation of trade equations based on producer rather than consumer behavior.

Both Burgess and Kohli focus on trade equations for intermediate goods. Goods are assumed to be inputs or outputs of the production sector, and there is no direct consumer demand for traded goods. Hence $D_i(\bullet) = 0$, so that $S_i(\bullet)$ represents import demand or export supply. In this case an import demand function should be derived from a technology or cost function rather than a utility function. The empirical argument for this approach is that trade in intermediate inputs represents the bulk of international trade, and, even when consumer goods are traded, they must go through some processing or retailing before final consumption.

Burgess (1974a, 1974b) derives and estimates two models of import demand in which firms are assumed to hire minimum cost combinations of imported and domestic inputs (capital and labor). In one model (1974a), investment and consumer goods are outputs, and in the other (1974b), a single output is produced. The technology is sufficiently general in both cases to allow a test of separability between imported inputs and domestic factors. This test is important since input separability would imply the popular specification of import demand in (4) is consistent with the cost minimization model of import demand (where prices are interpreted as input prices). U.S. data are used (1929-1969 is the period for (1974a) and 1947-68 for (1974b)), and in both cases, the separability hypothesis is rejected.

Using Canadian data for 1949-72, Kohli (1978) estimates a similar model. He estimates import demand and export supply simultaneously with domestic demand and supply of factors and outputs. A translog technology is assumed so that separability restrictions can be tested rather than assumed.

One of the most illustrative studies in this area is a theoretical one by Kohli (1982). He examines the implications of using different measures of domestic price and activity variables in estimating equations derived from a common structural model. Throughout the analysis, imports and a domestic composite factor are inputs to produce domestic gross output. The derived demand for imports and the domestic factor plus the unit cost of output are simultaneously determined for cases of constant and variable returns to scale. The natural estimating equation in this case would be

$$D_i^m = h\left(p_i^m, v_i, q_i\right) \qquad (6)$$

where v_i is the price of the domestic input and q_i is gross domestic output. Kohli shows that other estimating equations can be derived from the same structural model. For example, equation (4) can be derived from his model. The important point is that when (4) is used in lieu of (6) the price elasticity of demand for imports will be different. With constant returns to scale, the import price elasticity from (4) will be $[(\omega - \theta)/(1 - \theta)]$ where θ is the elasticity of substitution between imports and the domestic composite factor and ω is expenditure on the domestic factor as a share of expenditure on gross output. The import price elasticity from (6) will be $[(\omega - 1)\theta]$. Since $\omega<1$, the latter expression will always be negative, but the price elasticity from (4) may be positive. This property carries over to the case of variable returns to scale. Thus a positive estimate of price elasticity of demand in this model need not indicate rejection of the model. Thus Kohli's example shows how critical knowledge of the theoretical structure can be in interpreting elasticities.

ECONOMETRIC SPECIFICATION AND EVALUATION

In the previous section we focused on trade equations derived from alternative behavioral models. It is clear that different elasticity estimates may occur in the literature because of differences in behavioral models assumed. It is also clear that the same behavioral model can yield radically different elasticity values when different measures of variables are used. In the latter case, different estimates do not indicate rejection of a model, but call for care in interpretation. The tricky issue is when differences in elasticity estimates should make a researcher suspicious of a model (or class of models).

While the theory gives some guidance in this regard, it does not give a researcher sufficient tools to choose among alternative empirical estimates. Consider, for example, the intermediate goods model of Kohli (1982). Whether one estimates equation (4) or (6), the theory does not indicate the precise form of functions f or h, nor does it suggest whether lagged values should be included. While the theory can indicate how to interpret elasticity values in the two cases [i.e. (4) and (6)], it does not address issues related to errors in measurement of variables. Nor does theory give sufficient guidance on the appropriate procedure when data availability (or ignorance of the researcher specifying the model) leads to omission of an important variable. In this section we focus on the econometric literature dealing with such issues.

Dating at least to the classic works of Theil (1957) and Griliches (1957), economists have been aware of the deleterious effects of misspecification of regression models due to omission of relevant explanatory variables, use of an incorrect functional form, dependence between regressors and disturbances, etc.[10] Much of the attention to specification error has involved tests of whether observed regressors belong in some (presumably otherwise correctly specified) regression; use of simultaneous equation techniques (generally without first testing for the presence of endogenous regressors); ad hoc techniques such as consideration of sign and significance of regression coefficients and of R^2; etc. At least in applied work, methods regarding choice of competing models have been based largely on maximum R^2 and sign and significance of estimated coefficients. Recently, however, econometricians have become increasingly involved in methods of evaluation of regression models beyond the common ad hoc procedures. Both Bayesian and classical econometricians have developed statistical techniques for evaluating particular models and for choosing among competing models. We discuss and give examples of a number of these contributions, but first a few general comments about the econometric literature dealing with specification issues are in order. Since the econometric issues we discuss apply to any of the economic models of Section 2, we shall use general notation for linear regression models. It should also be noted that the term "model" in this section refers to a precise specification. This means, for example, that two different functional forms for a single behavioral "model" would be classified here as two competing models.

At the risk of over-simplification, we suggest that most specification problems in economic research can be represented by the following simple structure. Suppose a researcher posits a model of the following form:

$$Y = X\beta + \varepsilon \qquad (7)$$

where Y is a T x 1 observed vector of dependent variables, X is a T x k observed regressor matrix of rank k, β is a k x 1 vector of unobserved coefficients and ε is a T x 1 vector of unobserved random deviates. The researcher is interested in an element(s) of β. However, it is usually the case that an alternative model is a likely explanation of Y:

$$Y = X\beta + Z\lambda + \varepsilon \tag{8}$$

The variables in the matrix Z may be transformations of the variables in X, alternative measures of some variable (e.g., real national income versus gross domestic output), other economic variables, or any of a host of factors relevant to explaining variation in Y. In the fortuitous event that the elements of Z are well-specified and observable, there are few difficulties since the bulk of the testing literature deals precisely with such a framework.

However, numerous practical problems can arise in considering the alternative structure: variables may be unobservable, economic theory is typically very vague regarding possibilities for the alternative structure, inclusion of additional variables might exhaust available degrees of freedom, etc. Since interest centers on some element(s) of X and its associated coefficient(s), Z comprises a set of "nuisance" variables which should only be included to the extent that exclusion undermines the quality of estimators of β.[11]

A typical response to this problem is to estimate the first model, ignoring the possible relevance of the second, and if the estimated equation looks reasonable (coefficients have the right signs and are significant, the R^2 is high, etc.) the results are presented as if they are in some sense a true representation of the effects of X on Y. Another researcher might then criticize the first work because of the possible deleterious effect of omitting Z, alter the regression model by including some of the elements of Z (possibly excluding some of the elements of X) and, if the results look reasonable, present the results. All too often the economic implications change and readers are provided little guidance in choosing between studies. More information needs to be provided about the adequacy of models beyond conformance with a priori expectations.

Closely related to this problem of the possible existence of vague or ill-defined alternative models is the problem of choosing among competing, though well specified, alternative models which are "non-nested", meaning that none of the models can be derived from the others by the use of parametric restrictions. An alternative to model (7) might be written as

$$Y = W\gamma + \varepsilon \tag{8'}$$

where W is known and observable but its elements are not all found in the regressor matrix X nor are the elements of X all found within the matrix W. For example, model (7) might be an import demand function given by (4) in which case import price, other prices, and real national income are elements of the matrix X, and model (8') might explain imports in terms of the price vector and trend income.

There are several recent strands of (often closely related) econometric research germane to evaluating models such as (7) in the light of possible relevance of a vague or ill-defined alternative model such as model (8), and to choosing among competing non-nested models such as (7) and (8'). For ease of exposition, we discuss this research under four headings: goodness-of-fit, non-nested test procedures, specification searches, and specification error tests. The following is a general discussion, and Section 4 reports examples of each applied to international trade equations.

(A) Goodness-of-Fit Procedures. An early procedure for choosing among competing, though well-specified, econometric models was to select the model with the highest R^2, but this method naturally leads to models which have a large number of regressors. An alternative is the adjusted R^2 proposed by Theil (1961) and it has become a common model selection procedure. Critics note that maximizing adjusted R^2, as well as R^2, is implicitly justified by some loss function, but the loss function is not explicitly specified. Alternatives based on explicit loss functions are given by Mallows (1964), Akaike (1970, 1973) and Amemiya (1980). Their alternatives to the adjusted R^2 specify a loss function and then derive an estimable measure of that loss which is to be calculated for each of several possible alternative structures. The model which minimizes the specified loss is then chosen.

(B) Non-Nested Test Procedures. Implicit in any use of goodness-of-fit procedures in the assumption that one of the models is the true model since some model is always selected. Non-nested test procedures are an alternative approach which have as a possible outcome the rejection of all models under consideration. This alternative to minimization of loss functions for choosing among non-nested alternatives is based on classical model testing procedures and follows from early work in the statistics literature by Cox (1961, 1962). The idea is to consider each of the models in turn as the null model and one compares the actual performance of the alternative model(s) with the performance that could be expected if the null model were true. For example consider models (7) and

(8') as the competing models, and begin by considering model (7) as the null model. Model (8') is estimated first ignoring model (7) and then by assuming model (7) is the true model. The results are compared and if not statistically different we are unable to reject model (7) as the true structure, otherwise we reject model (7). We then repeat the exercise assuming model (8') is the true model and either accept or reject that model based on our estimates of model (7). This approach can lead to an acceptance of both models (implying that the data are unable to distinguish between the models), a rejection of both models (implying that a third, and as yet unspecified model, is the true model), or an acceptance of one of the models and rejection of the other. The ideas underlying the Cox procedures were first applied to regression models by Pesaran (1974) and Pesaran and Deaton (1978); excellent surveys are provided by MacKinnon (1983) and McAleer (1984). An example in international trade is Thursby and Thursby (1987).

(C) Specification searches. How sensitive are results to specification of the model? Much of the formal analysis of specification searches is done within a Bayesian framework and elegant treatments of the ideas can be found in Chamberlain and Leamer (1976), Leamer (1978) and Leamer and Leonard (1983). Cooley (1982) is an early empirical example of specification searches. The argument is made that all researchers have prior notions about parameters of an economic model and, regardless of whether a "classical" or "Bayesian" approach is taken, those prior notions affect the model actually estimated. Unlike formal Bayesian analysts, the proponents of specification searches doubt the efficacy of specifying fully a prior density to represent those prior notions. Rather, they advocate presentation of estimates derived according to different, though reasonable, priors.

Consider a simple example. A researcher posits a model

$$Y_t = \beta_1 X_{1t} + \beta_2 X_{2t} + \beta_3 X_{3t} + u_t.$$

The focus of interest is the coefficient of X_{1t}, a variable known to be an important determinant of Y_t. On the other hand, the researcher considers X_{2t} and X_{3t} to be doubtful variables in the sense that prior densities for β_2 and β_3 would give high probability to values at or close to zero. X_{2t} and X_{3t} are, for the purposes of the research at hand, a pair of "nuisance" variables in that they would be included in the regression only if inclusion improves the estimate of β_1. How does one proceed with the specification search? The simplest, though not the only, approach is to estimate β_1 subject to the constraints $\beta_2 = 0$, $\beta_3 = 0$ and $\beta_2 = \beta_3 = 0$ as well as unconstrained, and to present upper and lower bounds for the four estimates of β_1.

(D) Specification Error Tests. An alternative to specification searches is evaluation of a single model using a battery of formal tests of specification as well as possible ad hoc measures of fit. With specification error tests one is often concerned with testing a null model against an alternative that is only vaguely defined. A variety of procedures have been proposed for evaluation of models using specification error tests; recent discussions of methodology and alternative procedures can be found in Pagan (1984), Pagan and Hall (1983), Godfrey (1984), Breusch and Godfrey (1986), Davidson and MacKinnon (1985), and Thursby (1985).

Let the model to be evaluated be given by model (7) above and we can state the null and alternative hypotheses as $E(\varepsilon|X)=0$ and $E(\varepsilon|X)=\phi\neq 0$, respectively, where ϕ is an unobserved T x 1 vector whose nature is unknown to the researcher and ϕ is not orthogonal to X. Thus, under the null hypothesis, estimation of β in model (1) using, say, ordinary least squares gives unbiased estimates of β, whereas under the alternative hypothesis such estimates are biased. Tests of such hypotheses are "nonconstructive" in the sense of Goldfeld and Quandt (1972) or "general" in the sense of Ramsey (1974). A prominent example of such tests is the RESET procedure first proposed by Ramsey (1969) and later modified and extended by Ramsey and Schmidt (1976) and Thursby and Schmidt (1977). The RESET procedure is a standard F test of the significance of the δ estimates in the augmented regression

$$Y = X\beta + V\delta + u \qquad (9)$$

where V is a T x G matrix of rank G of test variables such as powers of the X variables. If $\hat{\delta}$ is the ordinary least squares estimator of δ in (9), then the power of the test follows from the fact that

$$E(\hat{\delta}) = (V'MV)^{-1}V'M\phi$$

where $M = I-X(X'X)^{-1}X'$. $E(\hat{\delta})$ is nonzero under general conditions if $\phi \neq 0$; hence it is not necessary that the researcher have a prior notion that V and ϕ are related (see Thursby and Schmidt (1977)).

The Hausman test procedure (Hausman (1978)) is an alternative to RESET which is of use when the researcher is able (or willing) to specify the nature of ϕ so that consistent estimators are known under both the null and alternative hypotheses. The Hausman test compares the estimator which is consistent and efficient under the null with an estimator which is always consistent (though inefficient). In general the test can be formulated in the added regressor framework of the RESET procedure (see, for example, Ruud (1984) and Davidson, Godfrey and MacKinnon (1985)).

These various strands of econometric research and methodology are often closely related and can often be fruitfully combined in a single study. For example, specification error tests might be used to eliminate a number of competing models with the remaining (accepted) models subjected to a specification search. We noted above that non-nested tests can be used to reject any or all models under consideration, and in that sense they are specification error tests.

EMPIRICAL APPLICATIONS

Elasticities and Alternative Specifications of Imports

In Thursby and Thursby (1984) we examined whether the simple, equation specifications of aggregate import demand frequently used in empirical studies were reliable in the sense that they would pass a variety of formal and informal specification tests. We examined a total of 324 estimating equations for each of five countries (Canada, Germany, Japan, United Kingdom and United States). We considered the nine basic models listed in Table 2.1, each of which is a variant of equation (4). With the exception of model (i), all models introduce dynamic behavior by including lagged values of either the dependent variable or independent variables. Again with the exception of model (i), all models are estimated both in their basic form and including a dummy variable to test for a shift in the demand function or nonconstancy of the income coefficient. The latter test follows from results of Stern, et. al. (1979) which suggest a structural change in United States import demand in 1972 related to income and nonprice factors. To allow for the possibility that the breakdown in the Bretton Woods system after late 1971 or the post-OPEC increase in oil prices could have shifted any country's import demand, we estimated models which allow a shift in 1972.1 (denoted by letter "a" after model number) and in 1974.1 (denoted by letter "b" after model number). For models iic and iid the shifts are permitted in the coefficient of income; whereas for all other models the shifts occur in the intercept term.

All models are estimated in linear and log-linear forms since previous studies had assumed one of these forms. In addition, we estimated each equation using two measures of the dependent variable: the import quantity index and the real value of imports relative to the price index of imports. Finally, we estimated each equation using three measures of the price of other goods: the implicit price deflator, the wholesale price index, and the consumer price index. The data are for aggregate quarterly imports for Canada (1957.1 - 1977.4), Germany (1960.1 - 1978.2), Japan (1957.4 - 1977.4), United Kingdom (1957.1 - 1977.3), and United States (1955.1 - 1978.1).[12]

Our model evaluation procedures begin with the RESET procedure using powers of the included regressors as test variables. If the RESET statistic is significant, the model is discarded as being misspecified. If the RESET statistic is insignificant, we apply choice procedures for nested models.[13] If some model is nested within a broader model, we test the coefficients of the variables in the broader model which are not included in the nested model. If significant, we then accept the broader model; otherwise, we accept the nested model. For example, each of the basic models is nested in the same model including a dummy variable. Hence if the coefficient of the dummy variable in any of the lettered models is significant (insignificant), the relevant basic model is rejected (accepted). Next we eliminate models with insignificant income coefficients and/or significant positive price coefficients. Finally, we eliminate equations with an adjusted R^2 less than .7.

Table 2.2 summarizes the results. Columns 2-4 list the percentage of specifications rejected by RESET, nesting, and all other rules (insignificant income coefficient, positive significant price coefficient, or adjusted $R^2 < .7$). The last column gives the percentage accepted after all rules are applied. Several general tendencies are evident. Model (i), the only model which does not incorporate dynamic behavior through lagged adjustment, is rejected for every country either by RESET or nesting (88% of these being rejected by RESET). The other striking result is that, by our criteria, including lagged values of the dependent variable appears to be more appropriate than lagged values of price and income.

While the above exercise yields a more manageable number of models than were originally specified, a question remains as to whether the elasticity estimates for the accepted models are different from those for rejected models. Table 2.3 presents the mean elasticities for four groups of models: (1) accepted, (2) rejected by RESET, (3) rejected by nesting, and (4) those rejected by either the income, price, or adjusted R^2 rules. We tested the hypothesis that the mean elasticity for the accepted models was equal to each of the other means using an analysis of variance framework. Whenever a mean elasticity is significantly different from that for accepted models, its level of significance is given in parentheses following the elasticity.

For three short-run income elasticities and two long-run income elasticities, the mean elasticity for accepted models was significantly different from those rejected by RESET. Moreover, for Germany and United States, short-run income elasticities for the accepted models were significantly different from all three other groups.

On the other hand, there are only two cases where the mean price elasticities for accepted models and models rejected by RESET are significantly different (the Japanese and United Kingdom short-run elasticities). For four countries the mean elasticity for those models rejected by the income, price, or adjusted R^2 rules differs from that of the accepted models.

Export Demand Elasticities

To illustrate use of non-nested tests and specification searches we consider a number of single equation specifications of United States wheat exports to Japan. We use annual data for the period 1960 - 1985 and the method of estimation is ordinary least squares (OLS). The models examined are all simple econometric models but are nonetheless models similar to many which have appeared in the empirical literature on export models (see, for example, Gardiner and Dixit (1986), and Gallagher, Lancaster, Bredahl, and Ryan (1981), Konandreas and Schmitz (1978)). Our intention is not to defend OLS or any particular model, rather we hold constant the data set and provide an example of the use of non-nested test procedures and specification searches.

We start with the regression model

$$M_t = \beta_0 + \beta_1\, Jp/CPI_t + \beta_2\, Inc_t + \beta_3\, Stks_t + \beta_4\, USpr_t + \beta_5\, Canpr_t + \beta_6\, Strike_t + \varepsilon_t \qquad (10)$$

where M = Japanese per capita imports of US western white #2

Jp = Japanese resale price set by Japanese Food Agency

CPI = Japanese Consumer Price Index

Inc = per capita real Japanese income

Stks = Japanese per capita beginning stocks + production - exports

USpr = real import price of US western white #2 in yen

Canpr = real import price of Canadian #1 western red spring in yen

Strike = variable to reflect US west coast dock strike activity.[14]

Japanese wheat imports are purchased by the Japanese Food Agency (JFA), a government monopoly, which resells to wholesalers at a fixed price. The resale price is generally set annually and is typically above the import price. While Japanese wholesalers face the resale price set by the JFA and not the United States import price, we nonetheless include the import price because of its possible effects on purchases by the JFA. Due to similar effects we also include the import price of Canadian wheat. Both USpr and Canpr, however, are considered "doubtful" or "unimportant" in the sense that any prior probability densities we might consider for their coefficients would give high probability to values at or close to zero. The variables Stks and Strike are also considered "doubtful" or "unimportant". The specification search, then, begins with consideration of the 16 regressions formed using all possible combinations of Stks, USpr, Canpr and Strike in conjunction with the variables Jp/CPI and Inc.

Several papers (see, for example, Murray and Ginman 1976) have argued that, while homogeneity restrictions may be appropriate for "micro" level import and export demand equations, in equations explaining aggregate flows imposition of such restrictions may lead to a deterioration in the quality of coefficient estimates. Hence we also consider the above set of regressions with Jp and CPI entering separately rather than as a ratio.

Price and income elasticities for the 32 models are given in Table 2.4. Note that the price elasticities vary over the interval (.002, -.918) and income elasticities vary over the interval (.107, .825).

Since the models with Jp and CPI entering separately are each a non-nested alternative to a model using the price ratio, the 16 pairs of alternative models can be compared using non-nested test procedures. The particular test procedure we apply is the JA test described as follows (see, for example, McAleer (1984)). Consider the two competing models

$$y = Z_1\theta_1 + u_1$$

and $$y = Z_2\theta_2 + u_2$$

used to explain the dependent variable y. The regressor matrix Z_1 is not nested in Z_2 nor is Z_2 nested in Z_1. (With respect to our wheat equation, Z_1 might be the regressor matrix consisting of the price ratio Jp/CPI and the variable Stks and Z_2 would be the regressor matrix with variables Jp, CPI and Stks.) Consider also the augmented regressions:

$$y = Z_1\theta_1 + \psi_1 B_2 B_1 y + u_1$$

and $$y = Z_2\theta_2 + \psi_2 B_1 B_2 y + u_2$$

where $B_i = Z_i(Z_i'Z_i)^{-1}Z_i$.

The JA test procedure consists of t-tests of the two null hypotheses $\psi_1=0$ and $\psi_2=0$. If $\psi_i=0$ is rejected, then we reject the model $y = Z_i\theta_i + u_i$. It is possible to accept both models or to reject both models as well as to accept one and reject the other. A rejection of both models implies that both models are incorrectly specified and a third (and unspecified) model is correct. Acceptance of both models simply means that the data are unable to distinguish between the models.

Using a ten percent significance level we are able to reject every model which uses the split-out prices Jp and CPI rather than the price ratio. Of the price ratio models we reject those models which include the variable Stks. The accepted models are indicated in Table 2.4 by underlining the price and income elasticities. For the set of accepted models the price elasticities vary over the narrow range (-.745, -.918) and the income elasticities vary over the narrow range (.107, .309). Thus the non-nested procedure rejects those models with low price elasticities and high income elasticities (as well as some models with high price and low income elasticities).

CONCLUDING REMARKS

Price and income elasticities of demand and supply for imports and exports vary by commodity, country, and time period. Estimates often vary dramatically even when comparisons are limited to studies of narrowly defined commodities exported or imported by a single country. In this survey we have focused on recent literature related to estimation and evaluation of trade elasticities. The approaches we discuss involve testing and evaluation procedures applied both intensively and extensively to models appropriate to the study of trade relationships, and reporting results from the entire exercise. Any hope of narrowing the range of estimated elasticities from trade equations must rely on taking such an approach.

In the first part of the paper we emphasized the need to more carefully specify the underlying economic framework so that estimates of trade model parameters can be more clearly understood. We then turned to a discussion of potential specification errors arising when a researcher's prior notions about the precise specification of a model are vague or ill-defined. Much recent attention has been paid to this problem by econometricians and we reviewed several of the major strands of research relevant to this problem. Finally, we considered several empirical examples which illustrate the potential benefits from a more exhaustive approach to model specification.

Table 2.1. Alternative Models of Import Demand

i.	$Y_t = f(P_t, I_t)$	vb.	$Y_t = f(P_{1t}, P_{2t}, I_t/IT_t, IT_t, Y_{t-1}, D74)$
ii.	$Y_t = f(P_t, I_t, Y_{t-1})$	vi.	$Y_t = f(P_t, I_t/IT_t, IT_t, Y_{t-1})$
iia.	$Y_t = f(P_t, I_t, Y_{t-1}, D72)$	via.	$Y_t = f(P_t, I_t/IT_t, IT_t, Y_{t-1}, D72)$
iib.	$Y_t = f(P_t, I_t, Y_{t-1}, D74)$	vib.	$Y_t = f(P_t, I_t/IT_t, IT_t, Y_{t-1}, D74)$
iic.	$Y_t = f(P_t, I_t, Y_{t-1}, D72{*}I_t)$	vii.	$Y_t = f(P_t, P_{t-1}, I_t, I_{t-1})$
iid.	$Y_t = f(P_t, I_t, Y_{t-1}, D74{*}I_t)$	viia.	$Y_t = f(P_t, P_{t-1}, I_t, I_{t-1}, D72)$
iii.	$Y_t = f(P_t{*}P_{t-1}, I_t{*}I_{t-1}, Y_{t-1})$	viib.	$Y_t = f(P_t, P_{t-1}, I_t, I_{t-1}, D74)$
iiia.	$Y_t = f(P_t{*}P_{t-1}, I_t{*}I_{t-1}, Y_{t-1}, D72)$	viii.	$Y_t = f(\text{Almon lag on } P_t \text{ and } I_t)$
iiib.	$Y_t = f(P_t{*}P_{t-1}, I_t{*}I_{t-1}, Y_{t-1}, D74)$	viiia.	$Y_t = f(\text{Almon lag on } P_t \text{ and } I_t, D72)$
iv.	$Y_t = f(P_{1t}, P_{2t}, I_t, Y_{t-1})$	viiib.	$Y_t = f(\text{Almon lag on } P_t \text{ and } I_t, D74)$
iva.	$Y_t = f(P_{1t}, P_{2t}, I_t, Y_{t-1}, D72)$	ix.	$Y_t = f(\text{Almon lag on } P_{1t}, P_{2t}, \text{ and } I_t)$
ivb.	$Y_t = f(P_{1t}, P_{2t}, I_t, Y_{t-1}, D74)$	ixa.	$Y_t = f(\text{Almon lag on } P_{1t}, P_{2t}, \text{ and } I_t, D72)$
v.	$Y_t = f(P_{1t}, P_{2t}, I_t/IT_t, IT_t, Y_{t-1})$	ixb.	$Y_t = f(\text{Almon lag on } P_{1t}, P_{2t}, \text{ and } I_t, D74)$
va.	$Y_t = f(P_{1t}, P_{2t}, I_t/IT_t, IT_t, Y_{t-1}, D72)$		

where Y = quantity of imports
I = real gross domestic product
P = price of imports relative to other goods
D = dummary variable beginning in indicated year
P_1 = price of imports
P_2 = price of other goods
IT = trend income

Sources: Table 1 of Thursby and Thursby (1984).

Table 2.2. Import Demand Model Outcomes

Model	Percentage Rejected by Rule RESET	Nested Test	All other	Percentage Accepted
		All Countries		
1	88	12	0	0
2	42	28	10	20
3	48	21	7	24
4	71	12	6	12
5	65	16	8	11
6	53	19	9	19
7	80	6	7	8
8	95	1	2	2
9	77	11	4	8
All Models	65	15	7	13
		Canada		
All Models	73	11	3	13
		Germany		
All Models	59	13	2	26
		Japan		
All Models	63	20	2	15
		United Kingdom		
All Models	53	18	24	5
		United States		
All Models	79	13	1	7

Sources: Table 2, Thursby and Thursby (1984).

Table 2.3. Import Demand Mean Elasticities

Model	Income Elasticities Short-run		Income Elasticities Long-run		Price Elasticities Short-run		Price Elasticities Long-run	
Canada								
1. Accepted	1.20		1.35		-0.19		-0.46	
2. RESET	0.96	(5)	1.42		-0.25		-0.50	
3. Nested	1.36		1.13		-0.02	(5)	-0.26	(5)
4. All other	1.66	(5)	1.00	(25)	0.29	(5)	0.04	(5)
Germany								
1. Accepted	0.98		1.59		-0.22		-0.30	
2. RESET	1.08	(15)	1.24	(5)	-0.24		-0.29	
3. Nested	1.12	(15)	1.48		-0.24		-0.23	(25)
4. All other	1.65	(5)	1.37		-0.42	(5)	-0.37	
Japan								
1. Accepted	0.76		1.17		-0.17		-0.33	
2. RESET	0.84		1.15		-0.10	(10)	-0.40	
3. Nested	0.84		1.25		-0.13		-0.23	
4. All other	1.27	(5)	1.04		-0.08		-0.15	
United Kingdom								
1. Accepted	0.78		1.12		0.10		0.14	
2. RESET	0.75		0.97		0.17	(25)	0.18	
3. Nested	0.72		1.04		0.19	(20)	0.24	(15)
4. All other	0.62		0.90	(25)	0.03	(25)	0.05	(20)
United States								
1. Accepted	1.50		1.72		-0.04		-0.20	
2. RESET	0.77	(5)	1.39	(5)	0.00		-0.14	
3. Nested	1.84	(5)	1.83		-0.03		-0.16	
4. All other	2.36	(5)	1.81		0.24	(5)	0.19	(5)

Sources: Table 3 of Thursby and Thursby (1984).

Table 2.4. Wheat Export Elasticities

Independent Variables								Price	Income
Jp/CPI	Jp	CPI	Inc	Stks	USpr	Canpr	Strike	Elasticity	Elasticity
X			X					-0.745 *	0.221
X			X	X				-0.270	0.388
X			X		X			-0.780 *	0.302
X			X			X		-0.791 *	0.286
X			X				X	-0.887 *	0.107
X			X	X	X			-0.302	0.396
X			X	X		X		-0.307	0.396
X			X	X			X	-0.523 *	0.244
X			X		X	X		-0.763 *	0.309
X			X		X		X	-0.905 *	0.167
X			X			X	X	-0.918 *	0.159
X			X	X	X	X		-0.299	0.393
X			X	X	X		X	-0.544	0.249
X			X	X		X	X	-0.557 *	0.252
X			X		X	X	X	-0.911 *	0.165
X			X	X	X	X	X	-0.541	0.242
	X	X	X					-0.258	0.825 *
	X	X	X	X				0.002	0.259
	X	X	X		X			-0.447	0.794 *
	X	X	X			X		-0.418	0.817 *
	X	X	X				X	-0.620	0.635 *
	X	X	X	X	X			-0.003	0.259
	X	X	X	X		X		-0.007	0.257
	X	X	X	X			X	-0.339	0.165
	X	X	X		X	X		-0.456	0.773 *
	X	X	X		X		X	-0.759	0.615 *
	X	X	X			X	X	-0.749	0.631 *
	X	X	X	X	X	X		-0.012	0.241
	X	X	X	X	X		X	-0.327	0.164
	X	X	X	X		X	X	-0.331	0.164
	X	X	X		X	X	X	-0.757	0.622 *
	X	X	X	X	X	X	X	-0.325	0.169

* Coefficient associated elasticity significant at the 10% level.

Appendix: A General Model of Specification Error

In order to gain insight into specification error we consider a simple, though revealing, regression model subject to specification error. The model is a quite general regression model with two included regressors and an arbitrary number of omitted regressors. Our purpose is to see more clearly the relation between parameters of the misspecified models and mean square error (MSE) of estimators of included regressor coefficients.

Consider the model

$$Y_t = \beta_1 X_{1t} + \beta_2 X_{2t} + Z_t'\alpha + u_t \qquad t=1,\ldots,n \qquad \text{(A1)}$$

where the regressors X_{1t} and X_{2t} are scalars, Z_t is a column vector of regressors and t refers to the observational unit. For expositional ease define $X_{3t} \equiv Z_t'\alpha$. Without loss of generality all variables are assumed to have zero means. β_1, β_2 and α are composed of unknown regression coefficients and the u_t are independent and identically distributed with mean zero and variance σ_{uu}. For analytic ease we assume that Y_t, X_{1t}, X_{2t}, and X_{3t} are multivariate normal. The vectors $X_t = (X_{1t}, X_{2t}, X_{3t})'$ are independent of u_t and distributed identically across t with covariance matrix

$$\Sigma = \begin{bmatrix} \sigma_{11} & \sigma_{12} & \sigma_{13} \\ \sigma_{12} & \sigma_{22} & \sigma_{23} \\ \sigma_{13} & \sigma_{23} & \sigma_{33} \end{bmatrix}.$$

Let us suppose that the researcher is interested in the coefficient β_1 of X_{1t}. We shall compare the mean square error (MSE) of the ordinary least squares estimator of β_1 in the regression of Y_t on X_{1t}, X_{2t}, and X_{3t}($MSE(\hat{\beta}_1)$) with the MSE of β_1 in the regression in which the researcher erroneously omits X_{3t}($MSE(\tilde{\beta}_1)$).[15]

Based on results in Aigner (1974) and Kinal and Lahiri (1983), it is easy to show that

$$\mathrm{MSE}(\tilde{\beta}_1) = \frac{\sigma_{33}}{\sigma_{11}\left(1-\rho_{12}^2\right)^2} \cdot \Big\{\left(\rho_{13}-\rho_{12}\rho_{23}\right)^2$$

$$+\left[\sigma_{uu}\left(1-\rho_{12}^2\right)/\sigma_{33} + 1 + 2\rho_{12}\rho_{13}\rho_{23} - \rho_{12}^2 - \rho_{13}^2 - \rho_{23}^2\right]/(n-3)\Big\}$$

and

$$\mathrm{MSE}(\hat{\beta}_1) = \frac{\sigma_{uu}\left(1-\rho_{23}^2\right)}{\sigma_{11}(n-4)} \cdot$$

$$\left(1+2\rho_{12}\rho_{13}\rho_{23}-\rho_{12}^2-\rho_{13}^2-\rho_{23}^2\right)^{-1}$$

where ρ_{ij} is the simple correlation of X_{it} and X_{jt}. Note that for positive definiteness of Σ it is necessary that

$$1 + 2\rho_{12}\rho_{13}\rho_{23} - \rho_{12}^2 - \rho_{13}^2 - \rho_{23}^2 > 0. \qquad \text{(A2)}$$

As an indication of the potential problems with the omission of X_{3t} we calculate values of $\mathrm{MSE}(\hat{\beta}_1)$ and $\mathrm{MSE}(\tilde{\beta}_1)$ for the parameter values

ρ_{12}, ρ_{13} and ρ_{23} = .0, .5, .75, .9;

$\sigma_{uu} = \sigma_{11} = 1$;

$\sigma_{33} = 1$ and .2; and

$n = 25$ and 100

and results for MSE($\tilde{\beta}_1$) and MSE($\tilde{\beta}_1$)/MSE($\hat{\beta}_1$) are found in Tables 2.A1 and 2.A2. Two points of particular interest to emerge are (1) MSE's of the two estimators vary a great deal as parameters and estimators vary, and (2) the exclusion of X_{3t} can actually lead to an improvement in the estimation of β_1 (see, especially, results for σ_{33}=.2, n=25 and ρ_{23}=.9).[16]

Table 2.A1. Mean Square Errors - σ_{33} =1

	MSE($\tilde{\beta}$)				MSE($\tilde{\beta}$)/MSE($\hat{\beta}$)			
	ρ_{12}							
ρ_{13}	.0	.5	.75	.9	.0	.5	.75	.9
A. n = 25				ρ_{23} = .0				
.0	.09	.12	.21	.48	1.91	1.91	1.91	1.91
.5	.33	.55	1.45	*	5.19	5.73	5.73	*
.75	.63	1.08	*	*	5.77	4.24	*	*
.9	.86	*	*	*	3.45	*	*	*
				ρ_{23} = .5				
.0	.08	.21	.88	*	1.67	2.97	4.64	*
.5	.32	.21	.26	.48	4.45	2.97	2.27	1.90
.75	.62	.53	.88	2.80	3.24	4.64	4.64	4.11
.9	*	.82	1.56	*	*	3.23	2.29	*
				ρ_{23} = .75				
.0	.07	.33	*	*	1.37	2.93	*	*
.5	.3	.11	.17	1.15	2.74	1.70	1.52	2.91
.75	*	.33	.32	.49	*	2.93	2.44	1.83
.9	*	.55	.72	1.68	*	1.40	2.67	2.62
				ρ_{23} = .9				
.0	.05	*	*	*	1.14	*	*	*
.5	*	.08	.28	*	*	1.18	1.60	*
.75	*	*	.22	.15	*	*	1.30	1.30
.9	*	*	.38	.50	*	*	1.35	1.54

Table 2.A1. Continued.

	MSE($\tilde{\beta}$)				MSE($\tilde{\beta}$)/MSE($\hat{\beta}$)			
				ρ_{12}				
ρ_{13}	.0	.5	.75	.9	.0	.5	.75	.9
B. n = 100				ρ_{23} = .0				
.0	.02	.03	.05	.11	1.98	1.98	1.98	1.98
.5	.27	.47	1.34	*	19.30	22.43	24.12	*
.75	.58	1.02	*	*	24.25	18.31	*	*
.9	.82	*	*	*	15.00	*	*	*
				ρ_{23} = .5				
.0	.02	.13	.77	*	1.73	8.58	18.44	*
.5	.27	.13	.12	.16	16.99	8.58	4.88	2.93
.75	.57	.46	.77	2.56	13.79	18.56	18.44	17.22
.9	*	.77	1.47	*	*	13.75	9.85	*
				ρ_{23} = .75				
.0	.01	.27	*	*	1.42	10.99	*	*
.5	.26	.05	.05	.92	10.79	3.24	2.22	10.57
.75	*	.27	.22	.23	*	10.99	7.39	3.95
.9	*	.5	.62	1.47	*	5.81	10.59	10.45
				ρ_{23} = .9				
.0	.01	*	*	*	1.18	*	*	*
.5	*	.02	.19	*	*	1.47	4.94	*
.75	*	.17	.06	.16	*	4.63	2.24	2.68
.9	*	*	.29	.29	*	*	4.76	4.06

*Parameter combination gives negative definite covariance matrix.

Table 2.A2. Mean Square Errors - σ_{33} =.2

	MSE($\tilde{\beta}$)				MSE($\tilde{\beta}$)/MSE($\hat{\beta}$)			
				ρ_{12}				
ρ_{13}	.0	.5	.75	.9	.0	.5	.75	.9
A. n = 25				ρ_{23} = .0				
0	.05	.07	.12	.29	1.15	1.15	1.15	1.15
.5	.1	.16	.37	*	1.61	1.65	1.47	*
.75	.16	.26	*	*	1.49	1.04	*	*
.9	.21	*	*	*	.83	*	*	*
				ρ_{23} = .5				
0	.05	.09	.26	*	1.10	1.27	1.36	*
.50	.10	.09	.14	.29	1.40	1.27	1.18	1.13
.75	.16	.15	.26	.75	.84	1.35	1.36	1.10
.90	*	.21	.39	*	*	.84	.58	*
				ρ_{23} = .75				
0	.05	.11	*	*	1.04	1.02	*	*
.5	.1	.07	.12	.42	.87	1.07	1.05	1.06
.75	*	.11	.15	.29	*	1.02	1.11	1.08
.9	*	.16	.23	.53	*	.40	.84	.82
				ρ_{23} = .9				
0	.05	*	*	*	.99	*	*	*
.5	*	.06	.14	*	*	.99	.8	*
.75	*	.09	.11	.27	*	.54	.97	.96
.9	*	*	.16	.29	*	*	.57	.9

Table 2.A2. Continued.

	MSE($\tilde{\beta}$)				MSE($\tilde{\beta}$)/MSE($\hat{\beta}$)			
	ρ_{12}							
ρ_{13}	.0	.5	.75	.9	.0	.5	.75	.9
B. n = 100				ρ_{23} = .0				
0	.01	.02	.03	.07	1.19	1.19	1.19	1.19
.5	.06	.10	.29	*	4.45	5.01	5.16	*
.75	.12	.21	*	*	5.20	3.86	*	*
.9	.17	*	*	*	3.15	*	*	*
				ρ_{23} = .5				
0	.01	.04	.17	*	1.14	2.42	4.14	*
.5	.06	.04	.04	.08	3.93	2.42	1.73	1.36
.75	.12	.1	.17	.56	2.96	4.15	4.14	3.74
.9	*	.16	.31	*	*	2.95	2.1	*
				ρ_{23} = .75				
0	.01	.06	*	*	1.08	2.65	*	*
.5	.06	.02	.03	.23	2.50	1.40	1.22	2.61
.75	*	.06	.06	.09	*	2.65	2.13	1.53
.9	*	.11	.14	.34	*	1.29	2.44	2.40
				ρ_{23} = .9				
0	.01	*	*	*	1.03	*	*	*
.5	*	.02	.06	*	*	1.07	1.49	*
.75	*	.05	.03	.08	*	1.22	1.19	1.25
.9	*	*	.08	.10	*	*	1.26	1.43

*Parameter combination gives negative definite covariance matrix.

NOTES

1. We express appreciation to Walter Gardiner and Nancy Schwartz. This work was done while the authors were on leave at the University of Michigan.

2. Goldstein and Khan (1985) report that by 1957 there were 42 books and articles containing estimates of income and price elasticities of import and export equations. Stern et al. (1976) provide a bibliography of 130 studies, and Goldstein and Khan's reference list contains 84 studies dated after the comprehensive Stern et al. review.

3. For examples, the reader is referred to the surveys listed above.

4. As noted by Haynes and Stone (1983), all but 10 pages of Stern et al.'s classic (1976) 363 page book is devoted to demand equations.

5. Studies by Kravis and Lipsey (1978) and Isard (1977) cast doubt on the law of one price, which is one of the implications of the perfect substitutes model. The work of Richardson (1978) and Thursby, Grennes, and Johnson (1986) has similar implications for certain agricultural commodities.

6. For an exception see Gregory's (1971) study of demand pressure and United States imports.

7. See Houthakker and Magee (1969) for an early example of import demand and Thursby and Thursby (1984) and Goldstein and Khan (1985) for more recent examples. Much of the estimation of export demand has been done in an elasticity of substitution framework. The reader is referred to Goldstein and Khan (1985), Richardson (1973), and Leamer and Stern (1970) for discussion of this concept.

8. Notable studies of export supply are Goldstein and Khan (1978), Haynes and Stone (1983), Clark (1977), Dunlevy (1980), and Kohli (1978).

9. The typical specification of import demand implies that the utility or profit (cost) function from which demand is derived is separable. Illustrative articles in this regard are Winters (1984), Burgess (1974a), and Goldstein and Khan (1980). In this survey, we have abstracted from issues related to the aggregation of trade equations across supplying or demanding countries. See Armington (1969) and Winters (1984) regarding separability issues in this regard. Also see Grennes, Johnson, and Thursby (1978) and Johnson, Grennes, and Thursby (1979) for discussion with regard to modelling trade in agricultural commodities.

10. While general expressions for specification bias and inconsistency are well-known, few treatments of the problem give more than simple examples of how poor estimates can be. In the Appendix we present a general regression model, solve for the mean square errors of coefficient estimators both for correct and incorrect models, and then substitute values of the parameters of the model in a demonstration of the impact of specification error.

11. See the Appendix and, in particular, the result that inclusion of a relevant variable can increase the mean square error of coefficient estimators of other variables in the regression.

12. Data are from International Monetary Fund International Financial Statistics, and Organization for Economic Cooperation and Development Main Economic Indicators and National Accounts for OECD Countries. All data have been seasonally adjusted.

13. Ordinary least squares is used to calculate the RESET statistic. If the model is accepted we then test for the presence of ARMA processes among the disturbances and, if necessary, correct for the implied process before proceeding to the other model evaluation procedures. See Thursby (1981).

14. Data on Japanese production, beginning stocks, imports, exports and resale price are from the Foreign Agricultural Service, USDA. CPI, population and GNP are from International Monetary Fund International Financial Statistics. Import prices are from International Wheat Commission World Wheat Statistics. Strike activity is from Gallagher, Lancaster, Bredahl, and Ryan (1981). Updated by the International Longshoreman's and Warehousemen's Union.

15. This representation of specification error is very general and can refer to omitted variables (in an obvious way), incorrect functional form (the omitted term becomes the sum of second and higher order terms in a Taylor series expansion of the true function), endogenous regressors (the omitted term then represents that part of the true disturbance correlated with the regressors), etc.

16. See also Wallace (1984) and Leamer (1983).

REFERENCES

Aigner, D. J. "MSE Dominance of Least Squares with Errors-in-Variables," Journal of Econometrics 2(December 1974):365-372.

Akaike, H. "Statistical Predictor Identification," Annals of Institute of Statistical Mathematics 22(1970):203-217.

________. "Information Theory and an Extension of the Maximum Likelihood Principle," in B.N. Petrov and F. Csaki (eds.) Second International Symposium on Information Theory, Budapest: Akademiai Kiado, 1973.

Amemiya, T. "Selection of Regressors," International Economic Review 21(June 1980):331-354.

Armington, P. S. "A Theory of Demand for Products Distinguished by Place of Production," IMF Staff Papers 16(1969):159-176.

Berndt, E. R. and L. R. Christensen. "The Internal Structure of Functional Relationships: Separability, Substitution, and Aggregation," Review of Economics and Statistics 60(July 1973):403-410.

Breusch, T. S. and L. G. Godfrey. "Data Transformation Tests," Economic Journal (Supplement) 96(1986):47-58.

Burgess, D. F. "Production Theory and the Derived Demand for Imports," Journal of International Economics 4(l974b):103-117.

________. "A Cost Minimization Approach to Import Demand Equations," The Review of Economics and Statistics 56(May 1974):225-234.

Chamberlain, G. and E. E. Leamer. "Matrix Weighted Averages and Posterior Bounds," Journal of the Royal Statistical Society, Series B, 38(1976):73-84.

Cheng, H. S. "Statistical Estimates of Elasticities and Propensities in International Trade: A Survey of Published Studies," IMF Staff Papers 7(1959):107-158.

Clark, P. B. "The Effects of Recent Exchange Rate Changes on the U.S. Trade Balance," in P.B. Clark, D.E. Logue and R.J. Sweeney, eds., The Effects of Exchange Rate Adjustments, Washington: U.S. Treasury, 1977.

Clements, K. W. "A General Equilibrium Econometric Model of the Open Economy," International Economic Review 21(June 1980):469-488.

Cooley, T. "Specification Analysis with Discriminating Priors: An Application to the Concentration Profits Debate," Econometric Reviews 1(1982):97-128.

Cox, D. R. "Tests of Separate Families of Hypotheses," Proceedings of the Fourth Berkeley Symposium on Mathematical Statistics and Probability, 1(1961):105-123.

________. "Further Results on Tests of Separate Families of Hypotheses," Journal of the Royal Statistical Society, Series B, 24(1962):406-424.

Davidson, R. and J. G. MacKinnon. "The Interpretation of Test Statistics," Canadian Journal of Economics 18(February 1985):38-57.

Davidson, R., L. G. Godfrey, and J. G. MacKinnon. "A Simplified Version of the Differencing Test," International Economic Review 26(October 1985):639-648.

Dunlevy, J. A. "A Test of the Capacity Pressure Hypothesis within a Simultaneous Equations Model of Export Performance," Review of Economics and Statistics 62(February 1980):131-135.

Gallagher, P., M. Lancaster, M. B., and T. J. Ryan. The U.S. Wheat Economy in an International Setting: An Econometric Investigation, Technical Bulletin No. 1644, Economics and Statistics Service, U.S. Department of Agriculture, March 1981.

Gardiner, W. H. and P. M. Dixit. "Price Elasticity of Export Demand: Concepts and Estimates," Economic Research Service, U.S. Department of Agriculture, May 1986.

Godfrey, L. G. "On the Uses of Misspecification Checks and Tests of Non-Nested Hypotheses in Empirical Economics," The Economic Journal 94(Supplement 1984):69-81.

Goldfeld, S. M. and R. Quandt. Nonlinear Methods in Econometrics, Amsterdam: North-Holland, 1972.

Goldstein, M. and M. S. Khan. "The Supply and Demand for Exports: A Simultaneous Approach," Review of Economics and Statistics, 60(1978): 275-286.

________________________. "Income and Price Effects in Foreign Trade," in R.W. Jones and P.B. Kenen (eds.), Handbook of International Economics, Amsterdam: North-Holland, 1985.

Goldstein, M., M. S. Khan, and L. H. Officer. "Prices of Tradable and Nontradable Goods in the Demand for Total Imports," Review of Economics and Statistics 62(1980):190-199.

Green, H. A. J. Aggregation in Economic Analysis, Princeton University Press, 1964.

Gregory, R. "United States Imports and Internal Pressure of Demand," American Economic review 61(1971):28-47.

Grennes, T., P. R. Johnson and M. Thursby. The Economics of World Grain Trade, New York: Praeger Publishers, 1978.

Griliches, Z. "Specification Bias in Estimates of Production Functions," Journal of Farm Economics 39(1957):8-20.

Hausman, J. A. "Specification Tests in Econometrics," Econometrica 46(November 1978):1251-1271.

Haynes, S. E. and J. A. Stone. "Secular and Cyclical Responses of U.S. Trade to Income: An Evaluation of Traditional Models," The Review of Economics and Statistics 65(February 1983):87-95.

Houthakker, H. S. and S. P. Magee. "Income and Price Elasticities in World Trade," Review of Economics and Statistics 51(1969):111-125.

Isard, P. "How Far Can We Push the 'Law of One Price'?," American Economic Review 67(1977):942-948.

Johnson, P. R. "The Elasticity of Foreign Demand for U.S. Agricultural Products," American Journal of Agricultural Economics 59(1977):735-736.

___________. Thomas Grennes and Marie Thursby. "Trade Models with Differentiated Products," American Journal of Agricultural Economics 61(February 1979):120-127.

Kinal, T. and K. Lahiri. "Specification Error Analysis with Stochastic Regressors," Econometrica 51(July 1983):1209-1220.

Kohli, U. R. "A Gross National Product Function and Derived Demand for Imports and Supply of Exports," Canadian Journal of Economics 11(May 1978):167-182.

__________. "Relative Price Effects and the Demand for Imports," Canadian Journal of Economics (May 1982):205-219.

Konandreas, P. A. and A. Schmitz. "Welfare Implications of Grain Price Stabilization: Some Empirical Evidence for the United States," American Journal of Agricultural Economics 60(February 1978):74-84.

Kravis, I. B. and R. E. Lipsey. "Price Behavior in the Light of Balance of Payments Theories," Journal of International Economics 8(1978):193-246.

Kreinin, M. E. "Price Elasticities in International Trade," Review of Economics and Statistics 49(1967):510-516.

Leamer, E. E. Specification Searches: Ad Hoc Inference with Nonexperimental Data, New York: John Wiley, 1978.

__________. "Model choice and Specification Analysis," in Z. Griliches and M.D. Intriligator (eds.) Handbook of Econometrics 1(1983), Amsterdam: North-Holland.

Leamer, E. E. and H. Leonard. "Reporting the Fragility of Regression Estimates," The Review of Economics and Statistics 65(May 1983):306-317.

Leamer E. E. and R. M. Stern. Quantitative International Economics, Boston: Allyn and Bacon, 1970.

MacKinnon, J. G. "Model Specification Tests against Non-Nested Alternatives," Econometric Reviews 2(1983):85-110.

Magee, S. P. "Prices, Income and Foreign Trade: A Survey of recent Economic Studies," in P.B. Kenen (ed.) International Trade and Finance: Frontiers for Research, Cambridge: Cambridge University Press, 1975.

Mallows, C. L. "Choosing Variables in a Linear Regression: A Graphical Aid," presented at the Central Regional Meeting of the Institute of Mathematical Statistics, Manhattan, Kansas (May 1964).

McAleer, M. "Specification Tests for Separate Models: A Survey," Australian National University Working Paper No. 110 (October 1984).

Murray, T. and P. Ginman. "An Empirical Examination of the Traditional Aggregate Import Demand Model," Review of Economics and Statistics 58(1976):75-80.

Pagan, A. R. and A. D. Hall. "Diagnostic Tests as Residual Analysis," Econometric Review 2(1983):159-218.

Pagan, A. R. "Model Evaluation by Variable Addition," in D.F. Henry and K.F. Wallis (eds.) Econometrics and Quantitative Economics, Oxford: Blackwell, 1984.

Pesaran, M. H. "On the General Problem of Model Selection," Review of Economic Studies 41(1974):153-171.

Pesaran, M. H. and A. S. Deaton. "Testing Non-Nested Nonlinear Regression Models," Econometrica 46(1978):677-694.

Prais, S. J. "Econometric Research in International Trade: A Review," Kyklos 15(1962):560-577.

Ramsey, J. B. "Tests for Specification Errors in Classical Linear Least Squares Regression Analysis," Journal of the Royal Statistical Society, Series B, 31(1969):350-371.

__________. "Classical Model Selection Through Specification Error Tests," in Frontiers in Econometrics, ed. P. Zarembka, New York: Academic Press, 1974.

Ramsey, J. B. and P. Schmidt. "Some Further Results on the Use of OLS and BLUS Residuals in Specification Error Tests," Journal of the American Statistical Association 71(June 1976):389-390.

Richardson, J. D. "Beyond (But Back To?) the Elasticity of Substitution in International Trade," European Economic Review 4(1973):381-392.

__________. "Some Empirical Evidence on Commodity Arbitrage and the Law of One Price," Journal of International Economics 8(May 1978):341-352.

Ruud, P. "Tests of Specification in Econometrics," Econometric Reviews 3(1984):211-242.

Stern, R. M., J. Francis, and B. Schumacher. Price Elasticities in International Trade, An Annotated Bibliography, Toronto, MacMillan of Canada 1976.

Stern, R. M., C. F. Baum, and M. N. Greene. "Evidence of Structural Change in the Demand for Aggregate U.S. Imports and Exports," Journal of Political Economy 87(February 1979):179-192.

Taplin, G. "Models of World Trade," IMF Staff Papers, 14 (November 1967):433-455.

Theil, H. "Specification Errors and the Estimation of Economic Relationships," Review of the International Statistical Institute 25(1957):41-51.

_______. Economic Forecasts and Policy, 2nd Edition, Amsterdam: North-Holland, 1961.

Thompson, R. L. "A Survey of Recent U.S. Developments in International Agricultural Trade Models," Economic Research Service, U.S. Department of Agriculture, September 1981.

Thursby, J. G. "A Test Strategy for Discriminating between Autocorrelation and Misspecification in Regression Analysis," The Review of Economics and Statistics 63(February 1981):117-123.

__________. "The Relationship Between the Specification Error Tests of Hausman, Ramsey, and Chow," Journal of the American Statistical Association 80(December 1985):926-928.

Thursby, J. G. and M. C. Thursby. "How Reliable are Simple, Single Equation Specifications of Import Demand?," The Review of Economics and Statistics 66(February 1984):120-128.

__________________________. "Bilateral Trade Flows, the Linder Hypothesis, and Exchange Risk," The Review of Economics and Statistics 69(August 1987):488-495.

Thursby, J. G. and P. Schmidt. "Some Properties of Tests for Specification Error in a Linear Regression Model," Journal of the American Statistical Association 72(September 1977):635-641.

Thursby, M. C., T. Grennes, and P. R. Johnson. "The Law of One Price and the Modelling of Disaggregated Trade Flows," Economic Modelling (October 1986).

Yntema, T. O. A Mathematical Reformulation of the General Theory of International Trade, University of Chicago Press, 1932.

Wallace, T. D. "Efficiencies for Stepwise Regressions," Journal of the American Statistical Association 59(1964):1179-1182.

Winters, L. A. "Separability and the Specification of Foreign Trade Functions," Journal of International Economics 17(November 1984):239-264.

Woodland, A. D. International Trade and Resource Allocation, Amsterdam: North-Holland, 1982.

Chapter 3

Philip C. Abbott

Estimating U.S. Agricultural Export Demand Elasticities: Econometric and Economic Issues[1]

INTRODUCTION

Agricultural trade issues have emerged as a central concern in the current U.S. farm policy debate. Rapid expansion of agricultural exports during the 1970s spurred growth in production and allowed for rapidly increasing prices. Since the late 1970s, however, demand has slackened in the international market and prices have fallen. It is argued that the competitive position of the U.S. in international markets has declined, as well. Market shares for several commodities exported by the U.S. have fallen substantially, compounding the effects of the declining export market. Both the strong dollar and U.S. farm programs during the early 1980's are prominent among the reasons given for the decline in U.S. agricultural exports.

In the farm policy debate, much emphasis was placed on the trade effects of domestic policy proposals (Paarlberg, Webb, Morey, and Sharples). It has been argued, for example, that the U.S. support prices set floors for international prices. Competitors could capture market share by offering commodities at prices slightly below the U.S. loan rate. It is believed that lower U.S. price supports would stimulate demand and improve farm exports.

The importance of international market behavior is also seen in the debate on policies which effectively subsidize exports (Grigsby and Jabara; Abbott,1984; Abbott, Paarlberg and Sharples). Again, a change in the U.S. border price is seen by some as an important means of stimulating demand and so improving conditions in the farm sector. There is not unanimous agreement, however, on the consequences of these policies.

A key factor in the current policy debate is the price responsiveness of export demand faced by the U.S. agricultural sector. Export demand elasticities concisely summarize that price responsiveness. An export demand elasticity is defined as the percent change in demand for U.S. exports of a commodity in response to a one percent change in the U.S. border (export) price. It reflects the net behavior of both U.S. trading partners and competitors to U.S. border price changes. It allows determination of domestic (U.S.) supply and demand adjustments to policy changes through the use of a simple model of trade response to price adjustments. Hence, calculation of the costs and effectiveness of alternative proposals is facilitated through the use of these parameters.

State of the Art

Unfortunately, our state of knowledge on U.S. agricultural export demand elasticities is poor for most commodities. There is wide variability in estimates of these parameters. A variety of methods have been proposed to estimate these parameters. Differing methods have yielded substantially different results, however.

Gardiner and Dixit recently reviewed the literature on export demand elasticities. They report numerous inconsistencies in these estimated parameters and raise a number of econometric issues. Long run export demand elasticities for wheat were found to vary from 0.23 to 5.00, while coarse grains elasticities ranged from 0.41 to 10.18. Soybean elasticities ranged from 0.29 to 2.80. Similar variations were found for short run elasticities, with similar ranges found for these commodities.

The consensus of the studies also puts the total coarse grains export demand elasticity at around 1.5, while estimates for corn, which constitutes the major portion of coarse grains exports, are all less than 0.5. The magnitudes of estimates for other coarse grains are not sufficient to account for the large aggregate elasticity, nor is it reasonable to expect that they should. Other coarse grains cannot account for the bulk of adjustment to price -- their share of total exports is simply too small. Also, the high substitutibility of commodities within an aggregate classification (such as coarse grains) should lead to lower aggregate elasticities, with higher individual commodity elasticities, not the reverse.

Gardner and Dixit identify two distinct approaches to estimation of export demand elasticities -- direct estimation and synthetic estimation methods. Direct estimation involves regressions of U.S. agricultural exports on U.S. border prices. Synthetic methods, recognizing substantial econometric problems with the direct approach (Orcutt; Binkley and McKinzie), divide the behavior observed at the U.S. border into its component parts. A variety of methods exist for estimating the parameters describing those component behaviors and aggregating those behaviors into a net elasticity. Typically, individual country or regional behaviors are set and a competitive spatial equilibrium model is used to determine the aggregate effect of the several countries' price responsiveness at their borders. The principles underlying alternative synthetic methods are similar. The component economic behaviors included and the parameter values assumed are not.

Much of the variation in estimates of U. S. agricultural export demand elasticities derives from the method chosen. Direct estimation tends to yield relatively low elasticity estimates, whereas synthetic estimation yields high estimates. Consensus usually sets elasticities between the range of these two approaches. In the synthetic methods, somewhat lower estimates are generally found when account is taken of the policy interventions of governments (Bredahl, Collins and Meyers; Abbott,1979), although attention is usually limited to price interventions at an importer's border. Internal price interventions separating consumer and producer prices and the workings of non-competitive forces in international markets are generally ignored.

Organization

Econometric and economic issues concerning estimation of U.S agricultural export demand elasticities are discussed below. The more traditional econometric issues suggest that direct estimation of elasticities is likely to be biased toward zero. These issues are reviewed and methods which address these issues are examined. Proposed alternatives emphasize the use of constrained estimation methods, with synthetic estimation seen as an effective way of incorporating constraints arising from prior economic theory. If synthetic models are to be useful, however, economic issues related to international trade of agricultural commodities must be addressed. Relevant economic issues which cause trade models to deviate from straightforward supply-demand equilibrium frameworks are discussed. This paper concludes by looking at two recent attempts to implement synthetic models of international agricultural trade for the purpose of estimating agricultural export demand elasticities.

ECONOMETRIC ISSUES

A variety of traditional econometric issues can be identified which have limited our ability to estimate U.S. agricultural export demand elasticities. These issues have a more direct impact on direct estimation methods and will be used to argue against that alternative. They also affect synthetic estimation methods, and should influence approaches to estimation under that alternative. Specification error, identification error, multicollinearity and aggregation are examined below for both alternatives.

Specification Error

U.S. agricultural export demand is the net effect of the combined behavior of all trading partners and competitors. Factors affecting their behavior in turn affect U.S. export demand. All variables in the supply and demand models of importing or exporting countries are candidates for a regression model of U.S. net export demand. Furthermore, factors affecting agricultural policy, which in turn influence international price transmission, can also affect export demand. This compounds estimation problems by lengthening the list of relevant variables, many of which must be excluded from the specification. This results in potential bias of estimated parameters, as included variables capture the effects of excluded variables.

Specification error is of particular concern to the estimation of a single net export demand elasticity facing the U.S.. Direct econometric estimation of this function assumes that the true specification of the function is known and that the function is stable over the estimation (and simulation) period. However, the literature reviewed by Gardiner and Dixit includes radically different specifications. Given the large list of variables to choose from, and the need for simple specifications with a limited number of variables, this outcome is hardly surprising. Orcutt uses this argument to support his contention that estimates of trade elasticities are generally biased toward zero. Further, the changing nature of policy interventions over time or the exercise of market power by another country suggests that the function may not be stable.

While specification error is especially important for direct estimation, it is often encountered in estimation of individual country or regional models. Trade economists face the problem of estimating basic supply-demand models and incorporating the effects of policy interventions for a large number of countries. The size of this problem generally precludes careful consideration of issues and behaviors specific to individual countries or regions. Nearly identical model structures are typically assumed, with account taken of special characteristics of only the largest traders. The result is often poor estimates of parameters for those component behaviors in synthetic models.

One solution to this problem is to take specifications (or estimated parameters) from prior work on individual countries (or regions). A wealth of information exists with "country experts" which needs to be tapped by trade economists. Re-estimation of country models each time a trade model is created is impractical, guaranteeing overly simplistic specifications.

Identification Error

Identification error arises when economists wish to separately estimate supply and demand models, and the same variables are in the specification of each model. While both models require Q = f(P), supply relationships are generally direct relationships, while demand is characterized by inverse relationships. Data is observed for an equilibrium, and not for the separate relationships. Net import demand functions and net excess demand functions suffer from this same problem, as each is part of an equilibrium system for the international market.

When supply or demand is estimated for an agent who may be viewed as a small component of a market, competitive assumptions permit the use of standard econometric techniques (i.e. OLS). For larger agents, identification is a more serious issue. Hence, direct estimation of a net excess demand function, with one supplier and one demander, requires corrections for identification problems. Net import demand functions may be estimated using ordinary least squares regression analysis for small traders, but not for traders with market power.

Identification error is characterized by violation of assumptions on the error term in a regression. Assumptions on distributions of the error term for unbiased or consistent estimation are violated regularly in practice, due to identification error and for other reasons. While violations of these assumptions can be treated by modifications of estimation techniques, such modifications are rarely done.

Multicollinearity

Multicollinearity is a particularly difficult econometric issue to address. It arises when a strong linear correlation exists between explanatory variables in a regression (or between included and excluded variables). Given the structure of economic models, multicollinearity is a commonplace occurrence. Both supply and demand equations require prices of substitutes and complements in the specifications. When these effects are strong, equilibrium outcomes seldom permit prices of different commodities to stray far apart.

Multicollinearity has prevented successful direct estimation of the full matrix of own and cross price effects in most econometric studies. The bulk of the literature utilizes single commodity models and a single equation estimation approach. Typically, only a few significant cross price elasticities can be found. Magnitudes of estimates obtained are seldom intuitively plausible. As the commodities in question are generally close substitutes, widely varying price relatives for these commodities are seldom observed. Multicollinearity is an important limiting factor in data needed for both direct estimation of net U.S. export demand and for component behaviors. It is particularly troublesome for multi-commodity analysis.

Methods to address multicollinearity generally involve careful specification of economic models, and use of restricted functional forms to constrain estimation of those specifications. Where there has been success, multi-commodity models utilizing prior economic theory to limit the number of parameters and to utilize assumptions on the nature of substitution have been used (i.e., estimation of demand for energy by Hudson and Jorgenson). Imposition of constraints increases econometric efficiency when those constraints are valid. Tradeoffs exist, however, between the risks of introducing bias in estimation from overly restrictive constraints and the benefits from lowering the number of parameters estimated (while reducing the flexibility of a specification).

Aggregation

Aggregation of economic agents whose behaviors differ can also lead to biased estimates of parameters. Aggregation issues for the estimation of U.S. export demand elasticities must be faced at two levels. First, the degree of regional disaggregation must be set. Then, commodity definitions must be determined, involving aggregating potentially dissimilar products.

Regional Aggregation

Direct estimation of a net excess demand function at the U.S. border constitutes aggregation of all regions of a market into a single region, and aggregation of all agents within that region into a single, naive specification. To the extent that behaviors across regions differ -- and given the wide diversity of countries trading agricultural commodities, that is inevitable -- biases in estimation outcomes are certain. The specification error problem and this aggregation question are essentially one in the same issue. Aggregate variables are used as poor proxies for several differing regional variables.

Direct estimation is particularly prone to aggregation bias, as it constitutes complete aggregation of all regions. Binkley and McKinzie show using a Monte-Carlo study that this aggregation is inferior to separate estimation of supply and demand for the major traders in an international market. They argue that "analytical least squares," which involves aggregating regional behaviors using a competitive model of international trade, leads to more efficient estimation, as well as minimizing bias.

An issue faced in "analytical least squares," and in other synthetic estimation methods, concerns the number of regions specified. If countries are grouped together with dissimilar behaviors, problems similar to those encountered by direct estimation arise. The potential number of countries which could be treated is too great to practically address each separately. Typically, smaller countries are aggregated into regions on geographical bases.

This runs the risk, however, of introducing aggregation bias if dissimilar behaviors exist. It is not evident that geography necessarily represents the preferred basis for aggregation. Domestic behavioral parameters, such as income elasticities of demand, may be similar. Policies effecting net trade behavior can differ substantially across countries located near one another, however. Furthermore, when determining parameters to use for regional aggregates, or for that matter choosing variables in regressions of regional behaviors, single country data must often be used. We would be better off increasing the number of individual countries modeled, and lumping all excluded countries into a single rest-of-the-world category.

Commodity Aggregation

Commodity aggregation represents another problem which must be faced in estimating export demand elasticities. The basis for establishing commodity aggregation is generally prior expectations based on physical characteristics. This may lead one astray, however. For example, Patterson recently estimated substitution elasticities for wheat and flour in 23 wheat and 22 coarse grains import markets. While prior expectations are that wheat and flour are close substitutes, few large substitution elasticities and many incorrectly signed elasticities were found. It may be that problems associated with regional aggregation of some countries with excess milling capacity and other countries which must import flour lead to aggregation problems when these commodities are combined. Grennes, Johnson and Thursby's work on Armington models suggests that different classes of wheat may not be perfect substitutes, as well.

Unfortunately, when commodities are fully disaggregated, multicollinearity prevents successful estimation of well behaved substitution effects. What should be found is that commodities within an aggregate grouping are close substitutes. Hence, while trade volumes for the aggregate may be relatively inelastic, price elasticities for individual commodities within the group will be larger, reflecting this substitution possibility. As noted earlier, for coarse grains the opposite has been found in existing econometric work. This problem is compounded by inclusion of regions where commodities are close substitutes -- which insures prices remain together -- with regions where substitution is weaker.

These econometric issues have made direct estimation of the export demand elasticities faced by the U.S. problematic. They also account for some of the difficulties in the synthetic estimation approach (analytical least squares), where the supply and demand behavior of trading partners and competitors is estimated. Part of the difficulty encountered stems from looking at these issues in a single commodity framework. Approaches to solution of these problems, based on constrained estimation methods for multiple commodities, are considered below.

CONSTRAINED ESTIMATION

Resolutions of econometric problems during the past decade have followed two distinct paths. Time series analysts have concentrated on reduced form regressions. Structural conclusions and behavioral implications are precluded by this method. Of particular difficulty for trade and agricultural policy analysis is that separation of domestic and international impacts, and hence evaluation of trade related policies, is not possible. Identification of supply effects versus demand effects is also impossible with the reduced form procedure. The alternative has been to use economic theory, restricted functional forms, and constraints based on as much prior information as seems valid to increase estimation efficiency.

It is well recognized that imposition of valid (or nearly) valid constraints increases estimation efficiency. Constrained estimation is particularly effective in the face of multicollinearity. Invalid constraints can seriously bias results and reduce efficiency, however (Binkley and Abbott). Hence, good economic models of international markets, incorporating relevant behaviors and institutional structures, need to be specified (Abbott, 1986b).

Both direct estimation and analytical least squares constrain parameter estimates. The constraints for the latter correspond to equilibrium identities for a competitive transportation problem plus the economic structure of a supply-demand system. The exclusion restrictions which must be imposed implicitly by direct estimation do not have a sound theoretical base. Hence, the approach advocated here is to estimate separately domestic behavioral models for important countries or regions, and use a trade model to aggregate behaviors into the net demand faced at the U.S. border.

Synthetic Models and Elasticity Estimation

Sharples used a modified version of the "analytical least squares" approach to estimate the export demand elasticity for U.S. wheat. A desirable feature of the Sharples approach is that, where possible, prior estimates of behavioral parameters for individual countries were used. In fact, Sharples relied entirely on prior research and provides no new estimates of domestic behavioral parameters. The term "synthetic estimation" derives from the practice of using prior work rather than repeating econometric estimation for each component behavior specified.

Problems with that practice are greater for some parameters than others. Econometric evidence on domestic supply and final demand is quite good, and intuition has been developed which can aid judgement in choosing those parameters. Evidence on the transmission of international prices to domestic economies, especially if one believes partial, lagged adjustments occur or if the effects differ for consumer and producer prices, is not nearly as good. Similarly, stocks adjustment is not well understood. Most trade models simply use a reduced form net stocks model which is not well grounded in theory. Furthermore, while own price and income elasticities of demand may be well known, cross price effects, especially for feed demand, are not nearly as well understood. There is some evidence on elasticities of substitution in feed demand for Europe (McKinzie, Paarlberg, and Huerta), but the behavior for LDC's is not well known. The dynamics of increasing feed demand in LDC's is probably more important to the evolution of international agricultural trade, however.

Use of trade models to calculate net trade elasticities does not preclude estimation of the component behavioral models specified. What is more desirable, however, it to take advantage of good prior information, devoting resources to estimation of the crucial, poorly understood relationships. Prior information is also useful in choosing among model specifications and in establishing constraints to estimation.

Component Behaviors

The Sharples approach divides net trade behavior by individual countries into supply, demand and stocks adjustment. Since his is a short run model only, supply is fixed. It may be preferable to divide trade into five component parts, each of which will require either econometric estimation or use of prior estimates of behavioral parameters. Those components are:

1. Price transmission,
2. Food demand,
3. Feed demand,
4. Production (supply response), and
5. Stock adjustment.

Feed demand, food demand, and supply respond to domestic prices. The price transmission model establishes the linkage between domestic and international prices. Stock adjustment should be set consistent with the observed policy process as revealed in the price transmission model. Stock adjustment will be important only for the large exporters and for a few importers. The result is a net trade model for each country or region which summarizes the effects of international prices on trade, or:

$$M_j^k = f\left(P_j^k, \text{other factors}\right) \qquad (1)$$

where M_j^k is net imports of commodity k by importing country or region j, and P_j^k are border prices at point j. Implementation of these net trade models consists of combining the five component models.

Aggregation by World Models

Aggregation of individual country or regional behavior into net export demand faced by the U.S. employs a model of world markets. A spatial equilibrium model, like the model developed by Holland and Sharples, is typically used to represent a competitive international market. Elements of imperfect competition practiced by governments may be incorporated in the price transmission or stock adjustment models. The resulting net trade models may then be included in the spatial equilibrium model. The spatial equilibrium model finds the transportation cost minimizing pattern of trade, such that net trade in each country or region equals the border price determined trade levels, and equilibrium market clearing identities hold in the international market. U.S. export demand elasticities may be found by exogenously setting the U.S. border price or export supply and tracing out net exports for alternative settings (as in Sharples).

Restricted Functional Forms

In developing models of price transmission, feed demand and stock adjustment, structured behavioral models strongly based on prior economic theory, and which limit the number of parameters to be estimated, should be used. Also, for food demand and supply, those structured models can be used to take best advantage of existing information. Structured forms, such as CES functions, translog functions, and the linear expenditure system, help to insure theoretical validity of estimated models while minimizing the task of econometric estimation, by using constraints to reduce the number of parameters estimated and by imposing necessary relationships between parameters. Differing functional forms impose differing degrees of restriction, however.

A variety of approaches to demand estimation are available, including straightforward models incorporating price and income as arguments in a linear function, and those which add further structure as constraints to estimation by basing the specification on utility theory. Among the latter are models which impose few prior restrictions on parameters (e.g., the translog model) and those which impose strong prior restrictions (e.g., the linear expenditure system -- LES).

Food demand represented by an LES system obeys all theoretical restrictions necessary for a demand system. The LES is based on the prior assumption that all goods are gross substitutes, however. This system will miss any special substitution relationships among commodities. Hence, it may perform better as a specification for aggregate classifications (i.e. food grains and feed grains) than for individual commodities. Its advantage over the translog system is that limiting the number of parameters reduces multicollinearity problems. A translog model is more flexible in its ability to capture substitution relationships. But its lack of prior restrictions makes it more dependent on weak, collinear data sets to reveal those relationships.

A second advantage of structured models is that they may be easily benchmarked to a base equilibrium with a minimum of parameters assumed. With the only assumptions being income and own price elasticities, an LES yields a complete demand model fit to base year data. Translog models, requiring more parameters, are harder to benchmark.

An intermediate approach, more often used for supply, is the constant elasticity of substitution function. CES functions as initially proposed were specified for two input production functions. A general CES function for an arbitrary number of inputs does not exist. The alternative is to construct a set of nests of pairs of inputs. A constant substitution elasticity can be estimated for the inputs in each nest. All inputs in a nest have the same relative response to inputs from another nest. The structure of the derived demand for feed lends itself to this nesting structure, so that the imposed restrictions on substitution behavior need not be unreasonable.

For feed demand, the potentially large number of parameters needed to capture substitution effects can be reduced by viewing it as derived from the livestock (meat) production process. Use of a nested constant elasticity of substitution production function for meat can reduce estimation to a set of substitution elasticities without forcing identical elasticities for all inputs. Such a model will obey the theoretical restrictions of production and derived demand.

Prior restrictions which limit the number of parameters also limit the range of possible cross elasticities. That limitation is not nearly as severe for nested CES functions as for either Cobb-Douglas functions, where all elasticities of substitution are one, or for Leontief functions, which set substitution elasticities to zero.

Delphi Methods

Two basic techniques for determination of behavioral parameters of simulation models are direct econometric estimation and assumptions based on existing econometric evidence and judgements of experts. Econometric estimation involves collection of time series or cross-sectional data on all variables included in the specification of net trade functions for each trader in the model, or of all variables in each component behavioral model. Given the substantial work needed to implement this approach, the existence of substantial prior econometric work to draw upon, and the problems associated with direct econometric estimation, the second approach, a Delphi process, is often adopted to establish parameter values.

Delphi methods base parameter assumptions on prior econometric work and on expert judgement. Experience has shown that econometric estimates can be highly variable and unstable. For certain relationships, such as the demand for grain as a food, underlying economic behaviors are well understood. This prior knowledge limits econometric estimates reported. It permits relatively accurate judgements on parameters. For other behaviors crucial to trade performance, the information is not nearly so good. We can probably guess final demand elasticities reasonably well. The debate over U.S. export demand elasticities shows that a Delphi process aimed at setting that parameter directly is unlikely to come to a clear consensus. Price transmission elasticities were found in the Trade Embargo Study (McCalla, White and Clayton) to be more problematic than direct supply or demand elasticities, while little intuition existed to set cross-price elasticities of either supply or demand.

Where prior information is weak, a Delphi exercise is unlikely to succeed. Hence, our recommendation is to use existing information where experience and intuition is well developed, devoting resources to estimation of the poorly understood but important relationships.

ECONOMIC ISSUES

The strength of synthetic estimation rests in its use of economic theory to constrain model specification and estimation. Several issues may be identified which are important to behavior in international markets, and which cause deviations from the straightforward supply-demand framework typically employed. These issues include:

1. Rigidities in trade flow patterns,
2. International price transmission and government intervention,
3. Stock management and imperfectly competitive markets,
4. Feed use and structural changes in demand patterns, and
5. Supply irreversibilities and asset fixity.

Our belief is that the trade models specified to conduct synthetic estimation must address these issues if realistic excess demand elasticities are to be estimated. Often, the outcome is dictated by such prior assumptions (as ignoring these issues) rather than by the results of empirical estimation. Each of these issues, and implications for synthetic estimation, are examined below.

Trade Flow Rigidities

A problem with traditional synthetic approaches derives from trade flow rigidities. Observed trade flow patterns do not correspond to the specialized patterns produced by spatial equilibrium models. Nor do trade flow patterns adjust quickly to small price changes. Prices from different exporters are not equated at a country's border. Trade is frequently conducted through long established historical, cultural, and political interactions. Commodities vary by quality, nutritional and milling characteristics. Grennes, Johnson, and Thursby propose the use of Armington models to address this aspect of trade behavior. That model emphasizes importers' non-competitive behavior. Kolstad and Burris propose a model of imperfect competition for large exporters, based on conjectural variations, nested in the spatial equilibrium structure. Each of these will give less specialized, more realistic trade flow patterns. Responsiveness to changes in prices at an exporter's border is reduced by either approach.

The Armington model has been used several times to empirically analyze world grain trade (Grennes, Johnson, and Thursby; Honma and Heady; Figueroa; Abbott and Paarlberg; Patterson). Trade flows are treated as differentiated goods and are determined using a two-stage maximization process. Maximization of utility in the first stage determines the total level of imports of a commodity. Total imports are then allocated to specific exporting countries. Armington's approach may be characterized as a simple, restricted and ad hoc (but effective) means of capturing the rigidities apparent in observed trade flow patterns.

Exporter behavioral relations and price linkages are the same as in a competitive model. The modifications in the model occur only in importer behavior vis a vis sources of supply. Demand for each good is a function of price indices of goods and a country's total income. The second stage maximizes the utility obtained from each good subject to a source allocation constraint, yielding:

$$X_{ij}^{k} = X_{ij}^{k}\left(P_{ij}^{k}, \ldots, P_{mj}^{k}, M_{j}^{k}\right) \tag{2}$$

where X_{ij}^{k} is the trade flow for commodity k from exporter i to importer j. Hence, it provides an alternative model for explaining trade flow patterns consistent with an overall import demand level, M_{j}^{k}.

To simplify the model, Armington introduces two assumptions. The elasticities of substitution for the imperfectly substitutable commodities in each market are assumed to be constant and the elasticity of substitution between any two products in a particular market is identical to that between any other pair in that same market. These assumptions mean that constant elasticity of substitution (CES) functions are imposed on importers:

$$X_{ij}^{k} = M_{j}^{k}\, b_{ij}\left(\frac{P_{ij}^{k}}{P_{j}^{k}}\right)^{-\sigma_j} ; \qquad j = 1, \ldots, n \tag{3}$$

where σ_j is the elasticity of substitution in country j and b_{ij} is a country specific constant. Furthermore, under these assumptions the good price index, P_{j}^{k}, is merely the average price for imports paid by j:

$$P_{j}^{k} = \frac{\sum_{i} X_{ij}^{k} P_{ij}^{k}}{\sum_{i} X_{ij}^{k}} . \tag{4}$$

The Armington approach to estimation of trade shares is an example of the effective use of constrained models using restricted functional forms. Prices of several exporters at an importer's border are certain to be highly collinear. While the restrictions of the CES functions utilized are strong, without them too many collinear variables remain in the specification, and failure to obtain a well behaved model is virtually insured.

Results from estimation of U.S. export demand elasticities when account is taken for trade flow pattern rigidities will be presented later in this chapter. More experience is needed with models examining the exporters' role in determining sources of supply.

International Price Transmission

Government policy interventions in agricultural markets, changes in transportation or marketing cost structures, and lags in the adjustment of domestic market conditions to world price changes can limit the transmission of international prices. Most countries do not respond completely or instantaneously to such changes, so that the effects on trade are muted by that adjustment process. This effect was included as one of Sharples' component behaviors. Its importance merits further consideration.

The price transmission effect is recognized in the literature, but relatively little econometric work has addressed it (Abbott,1979; Jabara; Bredahl, Collins, and Meyers). A serious problem for econometric estimation is that changes in policy regimes will alter underlying relationship between domestic and international prices. Hence, the assumptions of a constant behavioral model underlying econometric estimation may not hold.

Government institutions in many countries control the bulk of agricultural trade or foreign exchange allocations for imports. Those countries are close but not identical to closed economies, at least in the short run. They may not, however, be able to resist adjustments to long run trends in international prices. The relationship between international prices and the domestic prices which actually bring about supply-demand adjustments internally in countries is crucial to the determination of importer and exporter responses to international price changes.

Bredahl, Collins, and Meyers proposed a methodology for incorporating this relationship between international and domestic prices. They define a price transmission elasticity, which is the percent change in a country's domestic price in response to a one percent change in the international price. The elasticity of net import demand or net export supply is adjusted by this parameter.

Unfortunately, they can identify only two possibilities with institutional evidence alone -- free trade or zero price transmission.

Abbott's results suggest that intermediate or partial price adjustments to international price movements are appropriate for several countries. Collins' short run estimates of these price transmission elasticities for wheat and corn indicate that there is indeed partial short run domestic price adjustments to world prices in many countries where government institutions intervene to control agricultural trade. The extreme cases used by Bredahl, Collins, and Meyers are inadequate to represent partial government adjustments to international price signals. Their cited institutional evidence does not insure that a country completely ignores the world market.

Liu and Roningen argue that regressions in which domestic supply and demand are related to international prices implicitly incorporate price transmission effects. They refer to the transmission effect in explaining the relatively low elasticities found in several instances. That procedure is effective if a relatively simple relationship exists between domestic and international prices. If other factors also impact on domestic prices or alter the relationship between domestic and international prices, then these net trade responses are likely to be biased downward -- international prices at the U.S. border may be poor proxies for domestic prices and for international prices at another country's border.

The method used by Liu and Roningen is limited to short run response. If some countries cannot resist pressures from rising (or falling) international prices in the long run, or if there is some lagged adjustment in the relationship between domestic policy and international prices, that economic behavior is simply lost in this approach. It also assumes a single relationship between domestic and international prices. Byerlee and Sain have recently shown that while consumer prices in LDC's are often subsidized, producer prices are as well -- but to a different degree. Deviations from world price levels may be less than previously assumed. The lack of good consumer price data has limited the estimation of price transmission to producer prices only. Yet short run response is in consumption, not supply. Hence, existing price transmission elasticities may overstate short run adjustments.

The specification of a model of the transmission of international price changes to domestic prices begins with the identity relating domestic and international prices. That identity determines a tariff (export tax) equivalent for government intervention in agricultural trade. Data limitations prevent direct estimation of tariff equivalents as a function of world price. World prices are generally reported at export points. While data on international shipping margins are available to some import locations, marketing margin information generally is not. An alternative is to imbed the function explaining tariff equivalents into the identity relating domestic and international prices, and use proxy variables to examine the influences of transportation costs and factors influencing policy formation.

Stock Management and Imperfect Competition

Stock adjustments affect trade flows and pose both conceptual and econometric difficulties for estimation of price elasticities. Stock adjustment is an important but often overlooked factor contributing to the net export demand faced by the U.S. In his synthetic estimation of the short run U.S. wheat export demand elasticity, Sharples found Canadian wheat stocks adjustment to be one of the major contributors to price responsiveness. The trade embargo study exercise verified this result.

In their study of international grain stocks, Sharples and Goodloe found that stocks for most countries are responsive to variability in production. They account for a significant fraction of the adjustment to production shortfalls. While stocks may be a small fraction of production, they are often a much larger fraction of trade. However, in a relatively simple model, Sharples and Goodloe found few significant relationships between stocks and the world price.

Zwart and Blandford argue that stocks policy and price policy are related, and that a model which attempts to explain stocks adjustment must account for the policy regime in place. They propose several models for alternative policy regimes and argue that a single general model (and by implication a single regression model of world price on stocks) is unlikely to reveal the underlying stocks-price relationship. The model needs to be tailored to the policy regime and the way in which stocks are used to complement price policy. While Sharples and Goodloe find no significant relationship between Canadian stocks and the world price in their simple model, they indicate that Spriggs and Lattimore and Zwart have found such relationships in more complicated specifications. Presumably, more carefully specified models of individual countries could lead to more instances in which world prices are found to affect stocks adjustments.

There are several motivations leading to stockpiling, and those motivations may lead to alternative relationships between world price and stocks adjustment. Private stocks may be held as working inventories or for speculative gains. Public stocks are held to complement domestic policies (i.e., the stocks which arise in the U.S. as a consequence of American farm policies), stocks which may be used to exercise market power (Paarlberg and Abbott argue this is one objective of Canadian wheat stocks), and food security stocks held as protection against high and variable import costs (as in many LDC's). For purposes of analysis, stocks held in private hands but which are held to meet policy requirements (i.e., the U.S. farmer owned reserve) may behave as public rather than private stocks.

Zwart and Blandford propose a simple modeling framework appropriate for private stocks adjustment. Those stocks are driven by the level of output -- which sets transactions demand for working stocks -- and by the domestic price stockholders face. A more complete specification could then add a model of price expectations and factors affecting the costs of holding stocks, such as interest rates or storage charges. Where data on private stocks are available, such models are likely to have already been estimated.

As Paarlberg and Abbott have argued for their model, the only part of the stocks adjustment model needed for the trade simulations is the response to price. Estimation requires a considerably more complete specification than does simulation. A problem for estimation is that domestic prices, not international prices, may drive private stockholding. Where international price linkages are weak, this private stocks adjustment is unlikely to be strongly affected by world prices.

For a small country balancing the costs of importing against the costs of holding stocks to achieve food security, stocks adjustment reduces to a derived demand for stocks which depends upon a stocks cost function and the world price. For a large exporter, the resulting model is a variation on that proposed by Paarlberg and Abbott, but with stocks adjustment and not price interventions as the policy instrument. The essential difference between the small and large country cases is that benefits accrue from manipulating the world price and must be accounted for in setting policy.

The advice of Blanford and Zwart, building separate models for each individual country tailored to the policy regime and circumstances in that country, is not practical for a large number of countries. Alternatively, one could specify the net cost function to the policymaker using standard measures, such as consumer and producer surplus and treasury costs. That faces two problems. First, the food security, or stability objective, is not represented easily in such a setup. Second, the policy regime has already been captured by the price transmission model.

If one assumes that the price transmission model captures price setting in response to world price changes, it can be shown that the above model reduces to requiring that the marginal benefit to holding an additional unit of stocks must equal the world price. That can be rearranged to yield a stocks adjustment function depending upon the world price and other factors determining the costs or benefits of holding stocks -- such as interest rates, storage charges, or the supply-demand conditions of a particular year. Direct estimation of that stocks adjustment model is then feasible, with arguments based on this logic. A more complete specification will hopefully yield more information on the price responsiveness of stocks adjustments than was found by Sharples and Goodloe.

Structural Changes in Demand

USDA supply and utilization data separate grain demand into demand for use as food directly and demand for use as a feedstuff. Feed demand ultimately is driven by demand for livestock products as food. While there has been considerable work on demand for grains and oilseeds, generally this distinction between uses for food and feed is not drawn.

Mellor and Johnson have argued that forecasts of demand for grain were drastically underestimated in the seventies because the rapid expansion of meat in LDC diets was not taken into account. Income elasticities of demand for grains as feed are greater than is usually assumed, due to the high income elasticity of demand for meat. The intuition upon which prior econometric work on agricultural demand is based holds more for food use than feed use.

In most LDC's, governments control trade. Diet composition is likely to be a consequence of policy. Income elasticities of demand are unlikely to capture the changes that policies undergo as diet improvement proceeds. One can identify transitional phases for those countries who have upgraded diets. Imports of meat come first. As domestic production capacity is built up, imports shift to feedstuffs. As diet improvement is achieved, growth in imports of feedstuffs slows. Hence, structural shifts in demand patterns need to be accounted for in econometric work.

Production Response and Supply Irreversibilities

Distinction between short and long run supply response to international price changes is another important problem. While it is typically assumed that agricultural supply (production) is fixed for periods less than a year, supply response in the long run is recognized to be strongly price responsive (Askari and Cummings). On the other hand, several countries may use stocks adjustments to alter trade in the short run, but may not wish to continually accumulate stocks over the long run. Hence, stocks adjustments should tend to diminish as a factor affecting trade in the long run. Policy also limits the extent to which short run price movements are reflected in domestic price changes, but that policy may ultimately have to respond to long run trends in international prices.

Production decisions must be made prior to full knowledge of market prices. Adjustment costs may be incurred if production patterns are to change, and those adjustments may only be made over a relatively long period of time, as either capital stocks are adjusted or as long run expectations are altered. Hence, both stochastic and dynamic elements are crucial to production price response. Short and long run behavior must be separately identified. C. Taylor has argued that "empirical application of duality to many stochastic dynamic problems is quite complex and may be more difficult than a primal approach," arguing against the use of some restricted functional forms proposed for other component behaviors.

Nerlove has developed an approach to agricultural supply modeling which has been extensively applied. That approach assumes that supply may be modeled as a lagged adjustment process, in order to capture expectations formation and adjustment costs. To limit the number of parameters which need to be estimated, Nerlove employs a Koyck lagged adjustment structure. To incorporate cross price effects in this model, researchers have simply included additional prices in the specification. The relationship between short and long run response is then assumed the same for all relevant prices. Cross elasticities are seldom well behaved when estimated using this framework.

A considerable literature exists in which Nerlove's approach has been applied. It is the most often used specification for agricultural supply modeling. This literature was reviewed by Askari and Cummings in 1976. The GOL model of Liu and Roningen also utilizes this structure. Own and cross price supply elasticities may be taken from these sources to implement the supply component of this model for a wide range of countries. Empirical results provide a strong basis for judgements on own price elasticities in countries or regions where prior econometric work is not available.

An important aspect of the underlying notions of supply behavior concern the irreversibilities associated with supply response. That is, as world prices increase, and are passed on to domestic producers, long run expectations are altered, and more importantly, investments are made to increase production. As prices decrease, those same investments are now sunk costs. Imports are unlikely to return to previously high levels on the down side of this cycle. The debate over adjustment costs and irreversibilities is important to trade economists and should affect how they incorporate long run adjustments into their models.

EXPERIENCE WITH APPROACHES

Two recent experiences in estimating U.S. excess demand elasticities for wheat and coarse grains are discussed below. A Delphi process was used in the USDA Trade Embargo study (McCalla, White, and Clayton) to set parameters for trade models. Export demand elasticities were derived using a competitive, spatial equilibrium models. This author's involvement in that activity led to several of the proposals in this paper. It also points to a caution on the application of this approach. In a follow up to that study by Patterson, the effects of adding an Armington model structure to the determination of trade flow patterns on trade elasticities was determined. That exercise illustrates the importance of assumptions on international market structure and performance.

Trade Embargo Study

Synthetic estimation of excess demand elasticities in connection with the trade embargo study employed two novel features. First, a Delphi process was used to set parameters of component behavioral models. Prior econometric estimates were assembled for the steering committee of that study, and decisions were made on demand price and income elasticities as well as price transmission and stocks adjustment elasticities. The elasticities decided on in that process are reported in Table 3.1. Second, theoretical restrictions on demand functions were used to generate a complete demand system for each region in the model. A linear expenditure demand system (LES) was fit to the base set of market outcomes using the own-price elasticity and one additional parameter--the Frisch parameter. Estimates of income elasticities of demand for both wheat and coarse grains were used to estimate a Frisch parameter for each country. This yielded a complete, theoretically consistent matrix of own and cross price elasticities for wheat and coarse grains. As the income effects dominate the substitution effects, the gross cross-price elasticities were negative and small. This outcome is characteristic of the LES and demonstrates the limitation of using an overly restrictive functional form.

Given the complete matrix of demand elasticities, the stocks adjustment elasticities, and the price transmission elasticities, trade elasticities for each region were calculated. Base supply-utilization data were used to benchmark behavioral models to 1980 market equilibria, and to aggregate these to net elasticities at the U.S. border. The trade elasticities were short-run elasticities, based on the assumption that demand and stocks adjust to current period price changes, while production does not. Given the set of excess demand and supply elasticities for all major traders, net U.S. export demand elasticities for wheat and coarse grains were derived. The results of those derivations are reported in Table 3.2.

These results suggest that export demand for both types of grain is quite elastic. The short-run excess demand elasticity for wheat was calculated to be 1.15, and that for coarse grains was calculated to be 1.65. This outcome proved to be quite controversial, both within the study team and elsewhere. Lines were drawn between those more familiar with direct estimation results, and those who prefer the synthetic methods.

Armington Model

One potential explanation of differences between direct estimation results and the results of the trade embargo study concerns the competitive nature of the international marketplace.

The assumptions of the Armington model discussed earlier suggest that trade flow patterns remain rigid --they do not immediately adjust such that importers buy from the cheapest source. Rather, historical patterns remain for a variety of reasons. As noted earlier, price responsiveness at a country's border is likely to be less in an Armington model than for a spatial equilibrium model.

As a follow-up to the trade embargo study, Patterson estimated elasticities of substitution of all regions included in the trade embargo study model. His results are presented in Table 3.3. Those substitution elasticities suggested a rather rigid market, consistent with casual observation of data on trade flow patterns in these markets. He used the remaining parameters as set by the Delphi process of the trade embargo study. Since collection of trade flow data was necessary to estimate the market share equations, and since that data differed somewhat from the data of the trade embargo study, replication of the spatial equilibrium results provided a somewhat different answer -- 0.78 for wheat and 1.40 for coarse grains. This illustrates the sensitivity of the synthetic approach to the weights used in aggregating differing countries to a net elasticity. It also points to stability problems for direct estimation. Other than trade flow and hence net trade data, Patterson's model replicated the same market conditions as the trade embargo study model.

As expected, Patterson's Armington model yielded a much lower set of net export demand elasticities -- 0.51 for wheat and 0.77 for coarse grains. While these are still higher than results of direct estimation, they explain a considerable portion of the discrepancy between these two methods. These results are similar to prior results by Honma and Heady for a linearized Armington model of wheat trade, also reported in Table 3.2. Patterson further shows that as the elasticities of substitution are increased, the Armington model duplicates results of the spatial equilibrium model. But that occurs for elasticities of substitution well in excess of 10, which are outside the range of estimation results.

Parameter Sensitivity

In the Delphi process carried out during the trade embargo study, it was recognized that several parameters were quite uncertain. Of greatest concern were the excess demand elasticities of the Soviet Union and the elasticities of stocks adjustment for Canada. The Soviet wheat excess demand elasticity was believed to fall in the range from -0.25 to -0.75. Canadian wheat stocks elasticity ranged from -0.4 to -1.0. Similarly wide ranges were set for coarse grains. When these parameters were set at the high side of the ranges, excess demand elasticities facing the United States increased to -1.328 for wheat and -1.745 for coarse grains. Similar declines in U.S. excess demand elasticities were found when these parameters were set at the low end of the established ranges.

The derived excess demand elasticities were also sensitive to the assumed price transmission elasticities. In fact, much of the differences in the calculation method results reported by Gardiner and Dixit rests in the treatment of transmission elasticities. The very large long run elasticities originally found by Johnson and Tweeten implicitly assumed that transmission elasticities equalled one --that domestic and international markets were fully linked. Bredahl, Meyers, and Collins using institutional evidence to argue that domestic and international prices were divorced, found much lower excess demand elasticities. Other recent calculation methods use assumptions similar to those of Bredahl, Meyers, and Collins.

Recent results which update Collins' estimates for several Latin American countries provide even greater transmission elasticities (U.S. Department of Agriculture, Western Hemisphere Branch). Those estimates plus recent estimates of staff members in IED/USDA served as the basis for the outcome of the trade embargo study Delphi process. Hence, importers were found to be more price responsive than in most of the previous synthetic methods, especially for coarse grains.

For countries which are sensitive to international price changes, the excess demand behavior of the spatial equilibrium model is also somewhat sensitive to assumptions on domestic demand and the stock adjustment parameters. For example, Eastern Europe was assumed to have a domestic wheat demand elasticity of -0.2. If that elasticity is changed to -0.1, a seemingly small change well within the error bounds of the Delphi process, the excess demand elasticity for wheat facing the United States changes from -1.15 to -1.11, or 4%. Since the margin of error (and the range of prior econometric evidence) on these parameters is substantial, the potential for error in the net price response faced by the U.S. is considerable. However, large changes in the estimated excess demand elasticity facing the United States (large enough to reduce these to the findings from direct econometric estimation) requires systematically changing many of the assumed parameters.

Work by Sharples and Patterson also show that synthetically derived elasticities are sensitive to base year data. That is, as the mix of imports changes from one year to the next, the weights used to aggregate to the net U.S. excess demand elasticity change and so do the elasticities. Patterson's replication of the synthetic estimates of the trade embargo study, using only slightly different base data, yielded elasticity estimates more than 15% lower for coarse grains and 25% lower for wheat.

More importantly, elasticities of substitution for trade shares have been found to be quite unstable, and intuition to set these parameters is poor at this time. Patterson's alternative specifications and stability tests produced widely varying estimates. Yet Patterson also shows that these estimates can significantly alter derived net excess demand elasticities. He concluded that the Armington model better represented small shocks, in the short run, but was too rigid for large shocks or longer run adjustments.

Armington elasticities and price transmission elasticities are likely to impact synthetic estimates the most. But our knowledge of these important parameters is the poorest. Stock adjustments parameters are also important, but poorly understood. More work is called for not only on estimation of these relationships, but also on better conceptualizing models of the underlying forces, so that more stable econometric relationships may be found.

CONCLUSIONS

The component behavioral models discussed above constitute a procedure for establishing net trade by individual countries. Those net trade models may be incorporated into a spatial equilibrium multi-commodity trade simulation model for grains which determines a competitive equilibrium. Spatial equilibrium models solved using a fixed point algorithm (as in Holland and Sharples) may incorporate theoretically derived behavioral models. One is not restricted to using linearized versions of those models. A second advantage is that the effects of transportation charges, and their effects on trade flow patterns, are captured. Armington models are believed to better capture this behavior in the short run, however.

Data requirements to econometrically estimate each of the component behavioral models are substantial. While considerable econometric work has been devoted to estimation of some of the component models, others have been ignored. It is felt that a good basis for establishing production (supply) and final demand models based on prior work exists. New econometric work should emphasize the price transmission model, relating domestic and world prices; public stocks adjustment, where public stocks are an important component of domestic policy and so affect trade flows; the derived demand for feedstuffs, where cross price effects are not well understood; and trade flow models improving upon the Armington structure.

Data requirements are less rigorous for simulation than for estimation. Hence, one advantage of synthetic methods following a Delphi process to set some parameters is that efficient allocation of research resources to problem areas may be applied. When prior econometric work may be exploited, the trade simulation model may be benchmarked to the observed equilibrium. As in Sharples synthetic estimation, that supply-utilization balance is most representative if an average of several years is assumed. The averaging process will insure that trade equilibrium conditions hold for the base.

The proposed synthetic estimation methods are well suited to examination of the U.S. agricultural export demand in a multi-commodity setting. The component behavioral models may be specified to enhance our ability to estimate cross price effects by utilizing, where possible, models derived from underlying microeconomic theory. Where prior parameter estimation is available, the proposed behavioral models allow determination of consistent own and cross price effects by exploiting underlying economic theory. The implications of domestic and trade policies are explicitly introduced through price transmission and stock adjustment models.

Overly restrictive assumptions can lead to unrealistic estimates, however. Both the Armington synthetic estimation results and the use of the LES structure to obtain cross elasticities in the trade embargo study demonstrate the power of prior assumptions to dictate results. Clever use of alternative specifications, based on an understanding of the markets involved, rather than use of less restricted specifications, is more likely to lead to successful results.

The weakest areas of prior information were identified as being estimates of price transmission, stock adjustment, trade flow rigitities, and substitution effects. It was argued that incorporation of more realistic underlying economic assumptions, both in estimation of these parameters, and in application of synthetic estimation methods, should lead to a better understanding of U.S.

agricultural export demand elasticities. Its advantage lies in the incorporation of behavioral models derived from duality theory and explicit treatment of endogenous policy interventions.

Table 3.1. Elasticity Assumptions from Trade Embargo Study Delphi Process

	Wheat				Coarse Grains			
Regions	Trans.	Price	Income	Stocks	Trans.	Price	Income	Stocks
U.S.	1.00	-0.20	0.10	-1.15	1.00	-0.30	0.25	-1.00
Canada	1.00	-0.20	0.10	-0.70	1.00	-0.45	0.35	-0.50
E.E.C.	0.10	-0.25	0.10	-0.50	0.10	-0.35	0.35	-0.50
W. Europe	0.25	-0.20	0.15	-0.50	0.70	-0.60	0.35	-0.70
Japan	0.10	-0.35	0.15	0.00	1.00	-0.50	0.30	0.00
Oceania	0.90	-0.10	0.10	-0.40	1.00	-0.30	0.20	0.00
S. Africa	0.05	-0.30	0.20	0.00	0.40	-0.40	0.20	0.00
E. Europe	0.40	-0.20	0.20	0.00	0.35	-0.30	0.35	-0.30
USSR	0.32	-0.20	0.20	0.00	0.38	-0.40	0.30	0.00
China	0.20	-0.45	0.35	0.00	0.00	-0.10	0.20	0.00
Mexico	0.54	-0.40	0.35	0.00	0.70	-0.50	0.40	0.00
Cen. Amer.	0.50	-0.40	0.35	0.00	0.30	-0.50	0.40	0.00
Brazil	0.20	-0.30	0.25	0.00	0.10	-0.60	0.40	0.00
Argentina	0.50	-0.30	0.25	0.00	1.00	-0.40	0.30	0.00
Venezuela	1.00	-0.30	0.25	0.00	0.60	-0.60	0.40	0.00
S. America	1.00	-0.20	0.25	0.00	0.60	-0.50	0.40	0.00
SubSahara	0.40	-0.45	0.30	0.00	0.30	-0.30	0.30	0.00
Nigeria	0.14	-0.20	0.30	0.00	0.20	-0.60	0.50	0.00
Egypt	0.25	-0.50	0.35	0.00	0.50	-0.30	0.30	0.00
N. Africa	0.40	-0.20	0.35	-0.50	0.90	-0.70	0.50	-1.00
India	0.01	-0.50	0.40	-1.20	0.00	-0.20	0.10	-0.60
S. Asia	0.20	-0.50	0.40	0.00	0.00	-0.20	0.20	0.00
Indonesia	0.40	-0.70	0.40	0.00	0.00	-0.50	0.40	0.00
Thailand	0.50	-0.40	0.35	0.00	0.50	-0.80	0.50	0.00
S.E. Asia	0.50	-0.40	0.35	0.00	0.50	-0.50	0.45	0.00
E. Asia	0.60	-0.50	0.40	0.00	0.60	-0.70	0.45	-0.40
Mid East	0.40	-0.20	0.35	-0.50	0.90	-0.70	0.50	-1.00
ROW	0.00	-0.70	0.40	0.00	0.00	-0.50	0.40	0.00

Source: McCalla, White, and Clayton

Table 3.2. Short Run U.S. Export Demand Elasticity Estimates

Source	Wheat	Coarse Grains
Trade Embargo Study	1.15	1.65
Patterson --		
Spatial Model	0.78	1.40
Armington Model (Estimated)	0.51	0.77
Honma and Heady		
Linearized Armington Model	0.44	----
Gardiner and Dixit*		
Calculation Method	2.50	3.30
Direct Estimation	0.15	0.60

*Rough median estimated or calculated values are reported from Gardiner and Dixit's survey article for comparison only.

Table 3.3. Elasticities of Substitution--Patterson's Estimates

Regions	Wheat	Coarse Grains
E.E.C.	2.29	0.71
W. Europe	4.70	2.19
Japan	0.00	2.11
E. Europe	2.08	0.84
USSR	3.67	0.15
China	0.00	0.98
Mexico	0.24	2.20
Cen. America	1.03	1.65
Brazil	0.80	1.48
Venezuela	0.00	3.77
S. America	2.31	0.67
SubSahara	1.67	1.74
Nigeria	1.62	0.00
Egypt	1.55	3.75
N. Africa	3.20	3.22
India	3.99	0.00
S. Asia	0.77	0.00
Indonesia	1.40	0.00
S. E. Asia	0.00	4.14
E. Asia	2.59	3.75
Mid East	0.86	0.73

Source: Patterson

NOTES

1. Paul Patterson and Masaaki Kaneda provided valuable research assistance to this effort. Thanks are also due to Marshal Martin and Jerry Sharples for helpful comments on an earlier draft. This research was supported by a cooperative agreement with the Trade Policy Branch, International Economics Division, USDA. Responsibility for the views expressed in this paper rests solely with the author.

REFERENCES

Abbott, P. C. "A Methodology for Estimating U.S. Grain Export Demand Elasticities," Indiana Agricultural Experiment Station Journal Paper No. 10874, West Lafayette, IN, February 1986a.

_________. "Economic and Institutional Structure of Agricultural Trade: Developing Countries," presented at the AAEA Workshop on Modeling for Analysis of International Trade, Reno, NV, July 1986b.

_________. "Modeling International Grain Trade with Government Controlled Markets," American Journal of Agricultural Economics 61(1979):22-31.

_________. "U.S. Agricultural Export Promotion Activities: An Evaluation and Analysis of Options for U.S. Wheat Exports," AEI Occasional Paper, Washington, DC, November 1984.

Abbott, P. C. and P. L. Paarlberg. Impacts of the 1980 Suspension of U.S. Grain Sales to the U.S.S.R., Purdue University Agricultural Experiment Station Bulletin No. 504, West Lafayette, IN, October 1986.

Armington, P. S. "A Theory of Demand for Products Distinguished by Place of Production," IMF Staff Papers 16(1969):159-78.

Askari, H. and J. T. Cummings. Agricultural Supply Response: A Survey of the Econometric Evidence, Praeger Publishing, New York, 1976.

Binkley, J. K. and P. C. Abbott. "The Fixed X Assumption in Econometrics --Can the Textbooks be Trusted?" forthcoming in The American Statistician 41(1988).

Binkley, J. K. and L. McKinzie. "Alternative Methods of Estimating Export Demand: A Monte Carlo Comparison," Canadian Journal of Agricultural Economics 29(1981):187-202.

Bredahl, M. E., W. H. Meyers, and K. J. Collins. "The Elasticity of Foreign Demand for U.S. Agricultural Products: The Importance of the Price Transmission Elasticity," American Journal of Agricultural Economics 61(1979):58-63.

Byerlee, D. and G. Sain. "Food Pricing Policy in Developing Countries: Bias Against Agriculture or for Urban Consumers?" American Journal of Agricultural Economics 68(1986):961-69.

Collins, H. C. "Price and Exchange Rate Transmission," Agricultural Economics Research 32(1980):50-55.

Figueroa, E. "The Impacts of Movements in U.S. Exchange Rates on Commodity Trade Patterns and Composition." Ph.D. thesis, University of California-Davis, December 1986.

Frisch, R. "A Complete Scheme for Computing All Direct and Cross Demand Elasticities in a Model with Many Sectors," Econometrica 27(1969):177-196.

Gardiner, W. H. and P. Dixit. "The Price Elasticity of Export Demand: Concepts and Estimates," ERS Staff Report No. AGES860408, U.S. Department of Agriculture, Economic Research Service, May 1986.

Grennes, T., P. R. Johnson, and M. Thursby. The Economics of World Grain Trade, New York: Praeger Publishers, 1978.

Holland, F. D. and J. A. Sharples. "World Wheat Trade: Implications for U.S. Exports." Staff Paper No. 84-20, Department of Agricultural Economics, Purdue University. West Lafayette, IN, November 1984.

Honma, M. and E. O. Heady. "An Econometric Model for International Wheat Trade: Exports, Imports and Trade Flows," CARD Report 124, Center for Agricultural and Rural Development, Ames Iowa, 1984.

Hudson, E. A. and D. W. Jorgenson. "U.S. Energy Policy and Economic Growth, 1975-2000," Bell Journal of Economic Management Science 5(1974):461-514.

Jabara, C. "Trade Restrictions in International Grain and Oilseeds Markets," Foreign Agricultural Economic Report No. 162, ESCS, USDA, 1981.

Johnson, P. R. "The Elasticity of Foreign Demand for U.S. Agricultural Products," American Journal of Agricultural Economics 59(1977):735-36.

Kolstad, C. D. and A. E. Burris. "Imperfectly Competitive Equilibria in International Commodity Markets," American Journal of Agricultural Economics 68(1):27-36, February 1986.

Lattimore, R. and A. C. Zwart. "Medium Term World Wheat Forecasting Model," Commodity Forecasting Models for Canadian Agriculture, Vol. 2, Publication No.-78/3, Agriculture Canada, December 1978.

Liu, K. and V. O. Roningen. "The World Grain - Oilseeds - Livestock (GOL) Model, A Simplified Version," Staff Report No. AGES850128, IED/ERS/USDA. Washington, DC, February 1985.

McCalla, A., T. K. White and K. Clayton. Embargoes, Surplus Disposal, and U.S. Agriculture, USDA Agricultural Economic Report No. 564, Washington, DC, December 1986.

McKinzie, L. D., P. L. Paarlberg, and I. P. Huerta. "Estimating a Complete Matrix of Demand Elasticities for Feed Components Using Pseudo Data: A Case Study of Dutch Compound Feeds," European Review of Agricultural Economics, forthcoming, 1986.

MacKinnon, J. G. "An Algorithm for the Generalized Transportation Problem," Regional Science and Urban Economics 5(1975):445-464.

Mellor, J. W. and B. F. Johnson. "The World Food Equation: Interrelations Among Development, Employment and Food Consumption," Journal of Economic Literature 12(1984):531-74.

Nelson, C. and B. McCarl. "Including Imperfect Competition in Spatial Equilibrium Models," Canadian Journal of Agricultural Economics 32(1984):55-69.

Nerlove, M. The Dynamics of Supply: Estimation of Farmer's Response to Price. Baltimore: Johns Hopkins Press, 1958.

Orcutt, G. H. "Measurement of Price Elasticities in International Trade," Review of Economics and Statistics 32(1950):113-32.

Paarlberg, P. L. and P. C. Abbott. "Oligopolistic Behavior by Public Agencies in International Markets: The Case of World Wheat Trade," American Journal of Agricultural Economics 68(August 1986):528-42.

Paarlberg, P. L., A. J. Webb, A. Morey and J.A. Sharples. "Impacts of Policy on U.S. Agricultural Trade," ERS Staff Report No. AGES840802, IED/ERS/USDA, Washington, DC, December 1984.

Patterson, P. "An Application of the Armington Trade Flow Model in International Wheat and Coarse Grains Markets," M.S. Thesis, Purdue University, W. Lafayette, IN, August 1987.

Sarris, A. "Empirical Models of International Trade in Agriculture," in McCalla, A. and T. Josling, editors, Imperfect Markets in Agricultural Trade, Allenheld, Osmun, Montclair, NJ, 1981.

Sharples, J. A. "The Short Run Elasticity of Demand for U.S. Wheat Exports," IED Staff Report No. AGES820406, IED/ERS/USDA, Washington, DC, April 1982.

Sharples, J. A. and C. A. Goodloe. "Global Stocks of Grain: Implications for U.S. Policy," ERS Staff Report No. AGES840319, IED/ERS/USDA, Washington, DC, May 1984.

Spriggs, J. "An Econometric Analysis of Canadian Grains and Oilseeds," TB-1662, ERS/USDA. Washington, DC, December 1981.

Taylor, C. R. "Stochastic Dynamic Duality: Theory and Empirical Applicability," American Journal of Agricultural Economics 66(1984):351-357.

Taylor, L. Structuralist Macroeconomics: Applicable Models for the Third World, New York: Basic Books, 1980.

Takayama, T. and G. G. Judge. "Spatial Equilibrium and Quadratic Programming", Journal of Farm Economics 46(1964):67-93.

Thompson, R. L. "A Survey of Recent U.S. Developments in International Agricultural Trade Models," Bibliographies and Literature of Agriculture Number 21, ERS/USDA. Washington, DC, September 1981.

Tweeten, L. "The Demand for United States Farm Output," Food Res. Inst. Stud. 7(1967):343-69.

Zwart, A. C. and D. Blandford. "A Framework for Analyzing the Implications of Domestic Agricultural Policies for the Stability of International Trade," Cornell University Agricultural Experiment Station Paper No.--, A.E. Res. 85-14, Ithaca, NY, June 1985.

Chapter 4

John Dutton and Thomas Grennes

The Role of Exchange Rates in Trade Models

INTRODUCTION

The object of the paper is to analyze the effect of exchange rate changes on the volume of trade and prices. The use of elasticities to summarize the relationship between prices and quantities of traded goods is a common practice, and it can be a useful one. However, no single elasticity value can summarize the response of all product groups to all possible relative price changes in every possible economic environment. Every trade elasticity is conditional on many things, and failure to take into account differences in assumptions about underlying economic models can lead to misinterpretation of results. One reason for the wide range of trade elasticities reported in the empirical literature is that differences (sometimes implicit) in underlying assumptions make some of the parameter values incomparable with each other. A more careful analysis of model specification may narrow the range of plausible trade elasticities.

This paper will analyze the main issues related to trade elasticities, survey the earlier work, and present some new empirical results. The paper will emphasize the effect of exchange rates, although it is difficult to separate the effects of exchange rates from those of other prices. One interpretation of exchange rates in trade equations is that they measure the effects of omitted price variables. Section one will discuss some issues that are common to the effects of prices and exchange rates. Section two will survey the literature primarily related to exchange rates. Section three will consider theoretical issues and section four will present some empirical results. Although the present paper will discuss topics such as trade elasticities, econometric estimation, multi-country and multi-product models, and price transmission elasticities, they will not be emphasized because other papers in this volume will treat them in depth.

TRADE AND PRICE ELASTICITIES

The effect of prices and exchange rates on trade depends on several factors. Trade elasticities depend on the level of aggregation over commodities, countries, and currencies. They depend on the time period of the analysis, which might include overshooting or undershooting of prices relative to long-run equilibrium. The results depend on government policies influencing international price transmission, inventories and the behavior of the underlying macroeconomic variables. Trade elasticities also depend on methods of estimation and the nature of available data.

Aggregation Over Goods

Agricultural trade has been examined at various levels of product aggregation. Agricultural products have been treated as a single aggregate in studies designed to show the relationship between the agricultural sector and the rest of the economy (Deardorff and Stern; Clark, Houthakker and Magee; Gardner; Belongia). Intermediate levels of aggregation such as grain, meat, fats and oils, and dairy have been used. Narrowly defined products such as wheat, corn, and soybeans have been employed in other studies (Chambers and Just; Longmire and Morey), and these products have sometimes been distinguished by country of production (Armington; Grennes, Johnson, and Thursby). Because more substitutes exist for more narrowly defined products, trade elasticities increase as one moves from broader to narrower product categories. Aggregation is a convenient practice when dealing with certain macroeconomic issues, but it prevents one from analyzing problems involving changes in the relative prices of components.

When narrow product categories are used, one must decide how many cross-commodity effects to model explicitly. This is the familiar problem of partial versus general equilibrium. In general, the change in the quantity of wheat demanded in response to a change in the price of wheat depends on whether the prices of corn, soybeans, and other related products are held constant. The elasticity of demand for wheat will be greater if other prices are constant than if they adjust to clear all markets. More generally, the demand for wheat may also depend on prices of non-agricultural commodities (Adelman and Robinson), financial assets, factors of production (Seeley), and foreign variables (Anderson and Tyers). Even if data are available for all these variables, multicollinearity among them often makes stable and precise estimation of parameters quite difficult. Since all tractable empirical models omit some potentially relevant markets (or aggregate to levels that conceal important information), they are all partial to some degree. The importance of omitting some cross-commodity effects is unavoidably an empirical question that depends on the magnitudes of the relevant cross-price elasticities of demand and supply.

Some information about the importance of cross-commodity effects within the agricultural sector can be obtained by comparing the simulation results of a given model when cross-commodity effects are zero with the results when they take on plausible non-zero values. For example, the effects of currency appreciation on the prices and volume of exports of wheat, corn, and soybeans in the model of Longmire and Morey are shown in Table 4.1. In this model the introduction of "consensus" cross-price elasticities of demand and supply has a greater effect on exports than on domestic prices. The average price response is hardly affected, but the introduction of substitutability among the products makes the price responses more uniform. In the case of export volume, all changes become smaller.

Meilke compared the results of five multi-product models and a world wheat trade model which attempted to show the effects of a 5 percent decrease in production on prices and exports of wheat, coarse grains, and soybeans (see Table 4.2). The FAPRI model was run with and without cross-commodity effects, and price changes were reported to be sensitive to this relationship. However, with respect to the wheat price, the four other multi-product models produced results much closer to the world wheat trade model (which ignored cross-commodity effects) than the FAPRI results that included cross-commodity effects. Meilke has raised the question as to whether the export elasticity of demand for U.S. wheat might be greater than one when wheat is considered in isolation but less than one when other prices change and cross-commodity effects are considered.

Aggregation Over Countries and Currencies

Two-country models that aggregate the rest of the world into a single unit are useful for some purposes. They avoid the complexity of multi-country models, but they may suppress important country specific detail. Total exports can be explained but nothing can be said about the destination of exports by country. Also, country A may lower its tariff by 10 percent while country B raises its tariff by 10 percent. However, if the elasticities of demand for imports are different in A and B, imports in the rest of the world will change even though the average tariff level is constant. Country aggregation also affects elasticities directly, since the elasticity of demand for U.S. wheat is greater than the elasticity of demand facing all wheat exporting countries.

An infinite elasticity of demand facing one wheat farmer is consistent with a world market elasticity of -0.2. Similarly, an elasticity of less than one facing all wheat exporting countries collectively is consistent with an elasticity of greater than one facing the United States. For example, the price elasticities of export demand for wheat can be calculated for the five major exporting countries from the underlying elasticities of demand and supply. Assuming a demand elasticity of -.2 and a supply elasticity of +.2 for all countries, and using production and export data from 1985-1986, the resulting export demand elasticities are shown in Table 4.3 under the column heading, η_X^*. The export elasticities vary from -7.2 for the U.S. to -49.2 for Argentina, and they vary with each country's exports relative to the rest of the world's consumption and production. When all five major exporters are taken as a single aggregate, the export elasticity facing that unit is -1.8.

A kind of short-run elasticity of export demand can be obtained by setting the elasticity of export supply in the rest of the world equal to zero. The results are shown in the last column with the heading η_X^{**}. The results range from -3.7 for the United States to -24.7 facing Argentina. The calculated short-run elasticity facing the five exporters collectively is -1.0, which is not substantially different from the estimated elasticities obtained by Sharples and Paarlberg, Roe, Shane, and Vo, and Schmitz et al. These calculated elasticities may be interpreted as upper bounds, because they are based on the assumption that domestic price changes are completely transmitted to foreign markets. They also fail to take into account changes in prices of related products. However, the trade elasticities increase when exports decrease relative to world production, which has been happening in recent years. Thus, the demand elasticities facing each exporting country could increase even if each country maintained a constant share of world exports. Exporters compete not only against each other but also against producers in importing countries. In the case of wheat, exports were only 19 percent of world production in 1985-1986 (see Table 4.3).

Country aggregation is also related to market structure. Because the demand elasticity facing all exporting countries is less than the elasticity facing any single exporting country, countries have attempted to collude in international trade. In the words of Adam Smith "People of the same trade seldom meet together, even for merriment and diversion, but the conversation ends up in a conspiracy against the public, or in some contrivance to raise prices." The series of international wheat agreements since 1933 is an example of attempted collusion (Grennes, Johnson, and Thursby 1978) and there are current proposals to form a wheat cartel (Schmitz, et al). For many agricultural products it has become conventional to treat countries as the relevant exporting unit. This practice may result from the importance of state trading agencies (for example, marketing boards) in monopolizing the trade of individual countries. However, in the United States private firms compete in the export trade, and the elasticity of demand facing Cargill, for example, is greater than the elasticity facing all U.S. exporters (Caves and Rugel, M. Thursby). Thus, the aggregation issue applies to firms as well as countries.

Since country aggregation usually involves currency aggregation as well, many of the same issues arise for currencies. To represent the aggregate exchange rate either a single representative currency is used or some weighted average is computed. Problems arise when the differences in bilateral changes are large. When this occurs, differences in currency weights can produce quite different changes in aggregate exchange rates.

An example of the significance of currency aggregation is the experience of the floating dollar since 1980. On the average the dollar appreciated until March 1985 and depreciated since then.

However, the changes were quite uneven across countries, particularly in the latter period. If currencies are aggregated an optimal set of weights must be determined for various currencies, and different weighting schemes have produced significantly different results. Indeed, constructing effective exchange rate indices has become a growth industry (Pauls). For example, it makes a difference whether the dollar depreciates only against traditional importers of U.S. products or simultaneously against importers and competing exporting countries. Krissoff and Morey have constructed a three region model to show the difference between the two kinds of currency changes for the volume of U.S. exports of wheat, corn, and soybeans and their prices. The results are shown in Table 4.4. When the dollar depreciates by 10 percent against the importing country only, the U.S. wheat price rises by 6.5 percent and wheat exports rise by 5.2 percent. When the depreciation is against the exporting country only, the wheat price rises by 2.1 percent and exports increase by 1.5 percent. In the case of a uniform 10 percent depreciation against both countries, the wheat price rises by 8.9 percent and exports rise by 7.7 percent. In general, a uniform exchange rate change against all currencies magnifies the effects on prices and exports of all three products.

Time Lags and Dynamics

There is a consensus in the empirical trade literature that time lags are important in understanding the response of quantities to prices, and time lags may be longer for trade elasticities than for ordinary demand and supply elasticities (Goldstein and Khan, Thursby and Thursby). Since trade elasticities are excess demand elasticities, differences between long-run and short-run values are influenced by differences between long and short-run values of both domestic demand and supply elasticities. Lags in production depend on whether producers form their price expectations adaptively or rationally (Krissoff and Morey, pp. 7-12). Total supply response may be the same in both cases, but the time pattern of supply and export response may be quite different. It is common to formulate production as being fixed within a year and variable between years, but the effect of discontinuous production on supply is mitigated by the opportunity to sell out of inventories and trade with regions facing a different seasonal production pattern. The initial size of inventories can influence the time pattern of adjustment to an exchange rate change (Krissoff and Morey).

Insulating trade policies also influence the time path of trade elasticities. Since governments resist foreign price changes more in the short-run than the long-run, price transmission elasticities are greater in the long-run.

Econometric issues related to the estimation of time lags are surveyed by Goldstein and Khan (p. 1066-69). One issue is that time lags need not be the same for all explanatory variables and criteria must be established to determine the length of lags (Belongia). Also the estimated time lags may not be independent of the unit of observation. The same model may give quite different results depending on whether observations are monthly, quarterly or annual.

Speeds of price adjustment to exogenous disturbances may vary across markets. As a consequence, prices in certain markets may overshoot or undershoot their long-run equilibrium values (Stamoulis and Rausser; Obstfeld: Frankel). Since agricultural products are relatively homogeneous and traded in organized markets, they have been treated as having relatively flexible prices and being candidates for overshooting. Since overshooting implies a change in relative prices, it may alter production and consumption. However, suppose that over a month or a quarter prices overshoot their long-run equilibrium levels, but over a year they do not. If financial markets accurately reflect rational expectations, the economic consequences of overshooting for production, consumption, and trade may be small.

International Price Transmission

Governments frequently insulate domestic agricultural markets from changes in foreign prices and exchange rates. The elasticity of price transmission is designed to capture the effects of various policies that reduce the domestic response to external changes. In the simplest representation, the elasticity of export demand with respect to foreign prices (η_x^*) is proportional to the elasticity of demand with respect to domestic prices (η) and the elasticity of export supply by other countries (ε):

$$\eta_X^* = T\left(\frac{Q_R^d}{X}\eta - \frac{Q_R^s}{X}\varepsilon\right)$$

where T is the elasticity of price transmission and Q_R^d and Q_R^s are the quantities demanded and supplied by the rest of the world. In the extreme case of T = 0, the elasticity with respect to foreign prices will be zero no matter how responsive demand and supply are to domestic prices. Transmission elasticities may be different for producer and consumer prices and the relationship need not be linear (C. Collins; Roe, Shane, and Vo; Bredahl, Meyers, and Collins; C. Edwards).

Transmission elasticities are intended to be a summary measure of all government policies that separate foreign and domestic markets. Since they are indices of political behavior, they may be less stable over time than elasticities representing producer and consumer behavior. Empirical evidence shows considerable variation across countries and products (C. Collins; Roe, Shane, and Vo), which would be concealed by country and product aggregation. Like other elasticities, they are greater in the long-run than the short-run. For example, the EEC variable levy may offset nearly all foreign price variation within a year, but annual adjustments in target prices take into account world prices. Transmission elasticities may also be asymmetrical for price increases and decreases. For example, a decrease in the dollar price of grain requires a larger EEC export subsidy to keep European grain competitive. The budgetary implications may produce an asymmetrical policy response.

The simple formulation of price transmission focuses on products' own prices, but demand also depends on prices of related products. In particular, the domestic price of a product may be fixed, but if domestic prices of substitutes, complements, and inputs respond to foreign conditions, import demand will change. For example, if the world price of wheat falls, the wheat tariff in the EEC may be increased enough to hold the internal price of wheat constant. However, if the world price of soybeans falls, and soybeans are not subject to the levy, the quantity of wheat imports will decline. Attempts to insulate domestic agriculture from world conditions are complicated by the existence of imported inputs. In the case of meat, the quantity of imports depends on the price of feed grain. Cotton and wool can be imported as cloth or apparel as well as in raw form. As Americans have learned recently, a quota that keeps the price of raw sugar several times the world price does increase imports of products containing sugar. In some countries smuggling is quite responsive to differences between foreign and domestic prices.

The behavior of government agencies may make imports responsive to foreign prices even if domestic prices are controlled (Blandford). Managers of government inventories may increase purchases and imports when world prices are low and reduce imports when world prices are high. State trading agencies in countries with over-valued currencies may operate with a foreign exchange constraint which forces them to reduce imports when world prices increase (Scobie; Blandford).

When imports are imperfect substitutes for domestic products, estimation of price transmission elasticities becomes more complex (Blandford). Variation in price margins may also complicate estimation. Price transmission elasticities may themselves be a function of world prices.

The effect of alternative price transmission elasticities can be seen in Table 4.5 in terms of a model presented by Krissoff and Morey. In column one the price transmission elasticities are taken to be unity. In column two the transmission elasticities are taken to be 0.5 for wheat, 0.8 for corn, and 0.9 for soybeans. The effect is to reduce the response of all prices and quantities shown.

Export Supply, Inventory Demand, and Domestic Policy

In policy discussions of U.S. exports and in empirical work on the subject, the demand for exports has received more attention than the supply of exports (Goldstein and Khan). However, in general, the effect of a disturbance to the export market depends on both elasticities of export demand and supply. Following a devaluation, for example, the increase in export volume is an increasing function of the elasticity of export supply, and the increase in the domestic price is a decreasing function of the export supply elasticity.

The case of infinitely elastic export supply for particular products has received some attention in the agricultural economics literature (Mitchell and Duncan; C. Edwards). In the case of wheat, the United States has been treated as the residual supplier of wheat at a given price. Even though the private wheat sector is subject to increasing marginal cost, certain government policies could justify the assumption for a limited time period. To prevent world prices from falling, the government must be willing and able to add to inventories and reduce exports. To avoid rising world prices the government must possess sufficient inventories and be willing to add to export supply.

Economic policy could also produce the extreme opposite relationship. A fixed domestic support price combined with a variable export subsidy ("restitution") would bring about an export supply elasticity of zero. For example, European Community grain policy has made its export supply and import demand extremely inelastic in the short-run.

Since export supply is an excess supply, its elasticity (ε_X) depends on both domestic supply (ε) and domestic demand elasticities (η). It also depends on the elasticity of price transmission (T) between foreign and domestic markets. The relationship can be written as:

$$\varepsilon_x = T\left(\frac{Q^s}{X}\varepsilon - \frac{Q^d}{X}\eta\right)$$

During the 1970s when world prices were rising, it was common to assume that foreign prices were completely transmitted to the U.S. markets (T = 1). However, in the 1980s when prices have been falling, U.S. domestic policies have resisted the price decreases. For example, support prices have been binding for certain products for certain periods. Price transmission to the United States, which has been incomplete and asymmetrical, has altered the export supply elasticity.

Inventory demand can be separated from the rest of domestic demand, and government price support programs have been represented by altering the elasticity of inventory demand. Thus, binding price support policies indirectly alter export supply. In models that specify a separate inventory demand (Meilke; Longmire and Morey), much of the difference in trade elasticities can be attributed to differences in implicit assumptions about government inventory policy. Modelers have used inventory equations to introduce domestic policy in the same way they have used price transmission elasticities to represent the policies of foreign governments. Since these equations represent complex political interaction, it is not surprising that they are among the weaker components of economic models.

Even though there is little theoretical or empirical justification for particular parameter values, inventory demand elasticities play a prominent role in several multiple product models. In the six trade models compared by Meilke (see Table 4.2) the price response appears to be more sensitive to the elasticity of stock demand than to the elasticity of export demand, for a given exogenous reduction in supply. For all three products (wheat, coarse grain, and soybeans) the model with the least elastic inventory demand generates the largest price increase. For example, for wheat the FAPRI model generates a much larger price increase (17.1 percent) than the World Bank model (1.5 percent), even though the export demand elasticities are nearly identical (-0.3 and -0.4). The difference in price response appears to be attributable to the much smaller stock demand elasticity in the FAPRI model (-0.3 versus -5.9). In the case of soybeans both the Michigan State and the World Bank models contain the same extremely inelastic export demand (-0.1). The large difference in price response (20.8 versus 4.4) appears to be a result of different assumptions about inventory demand.

Domestic supply, the second component of export supply, may be less responsive to foreign prices because of insulating domestic policies. It may also be influenced by direct controls on production or input use. The tobacco program, which directly controls production, has the effect of reducing the elasticity of export supply. Programs that restrict land use also reduce export response. Since domestic programs have varied substantially across commodities, export supply elasticities may be quite different for products whose domestic demand and supply elasticities are equal. Products like soybeans, that have been relatively free of domestic restrictions, might be expected to have a more responsive export supply than wheat or corn. Complexity of individual commodity programs and changes over time contribute to the difficulty of isolating an export supply relationship.

Export subsidies are another factor influencing export supply. Subsidies have taken many forms including export enhancement programs, marketing loans, subsidized credit, and food aid. Eligibility for export subsidies has varied across commodities and importing countries. At times eligibility has been automatic, but at other times it has been at the discretion of the administration.

Many factors, including incomplete price transmission, inventory policy, production and land restrictions, and export subsidies, can influence the elasticity of export supply. At times these domestic policies may have made the actual export supply much less elastic than the excess supply of the United States. By restricting export supply these policies have reduced the response of exports to exogenous disturbances. Some portion of the unresponsiveness of U.S. exports that has been attributed to inelastic export demand may instead be a result of inelastic export supply. To the extent that these export supply considerations lead people to underestimate the true elasticity of export demand, policies of production and export control will appear to be more attractive to the United States than they actually are.

Estimation of Trade Elasticities

In his seminal 1950 paper, Orcutt analyzed the basic problems in estimating trade elasticities, and emphasized reasons why estimates would be biased downward. For example, goods with the lowest price elasticities within an aggregate show the greatest price variation.

In their recent survey of empirical trade, Goldstein and Khan concluded that most aggregate trade elasticities were greater than one in the long-run (two years or more, p. 1076). However, short-run elasticities are generally half as large and about 50 percent of the final adjustment occurs in one year (p. 1077). One implication is that the balance of trade may deteriorate following a price decrease before it improves. A second implication is that income effects may dominate price effects in the short-run.

For narrower product categories, price elasticities tend to be greater for manufactured products than for non-manufactures. Within the non-manufactured category, elasticities tend to be greater for raw materials and fuels than for food and beverages. Thus, aggregate trade elasticities may vary across countries because of (a) differences in commodity composition of trade and (b) differences in the importance of trade relative to total production and consumption in a country. The smallest trade elasticities should apply to relatively closed countries specializing in food and beverages.

Partly because of econometric problems, other methods have been used to obtain trade elasticities. Excess demand elasticities have been calculated from the underlying domestic demand and supply elasticities or, and they have been generated by simulations of multi-product models (W. Gardner and Dixit; Meilke). In other cases elasticities have been obtained by surveying informed observers (Abbott and Paarlberg).

Empirical supply elasticities tend to be greater for disaggregated products than for total exports, and time lags tend to be greater for export supply than export demand elasticities (Goldstein and Khan, p. 1087). Production lags are relevant for agricultural products, and supply elasticities can be influenced by production and acreage controls. Following devaluation, import prices tend to rise faster than export prices. The value of imports rises temporarily and the trade balance deteriorates temporarily (p. 1098). The precise result also depends on the aggregate demand policy carried out.

Macroeconomic Variables Held Constant

The effect of a price change on trade depends on what other variables are held constant. For example, the effect of a change in the price of wheat depends on whether other agricultural prices are held constant. Aggregating wheat with all other agricultural products implicitly holds constant relative prices within the aggregate and lets them move jointly against non-agricultural products. There will be a smaller change in the demand for wheat when all other agricultural prices vary in the same proportion than when they are constant.

The change in demand for a particular agricultural product also depends on whether prices of non-agricultural traded goods (80 percent of U.S. trade) are permitted to vary. Demand response also depends on factor markets, financial markets, and aggregate demand variables such as monetary and fiscal policy. The potential general equilibrium complications arising from interdependence among various markets has given rise to a distinction between partial and total elasticities (Dornbusch 1975). However, this terminology is confusing because every elasticity is partial to some degree in the sense of holding some potentially important variable constant. For example, a trade elasticity might be total in the sense of letting the relative prices of all commodities vary but partial in the sense of holding interest rates, the money supply, and real gross national product constant.

Dornbusch (1975) has emphasized that the effect of an exchange rate on prices and trade depends on the underlying monetary and fiscal policy. Since currency depreciation raises the domestic prices of traded goods, it makes a difference whether macroeconomic policy is used to hold the prices of non-traded goods constant or money gross national product constant. The latter case requires a more restrictive macro policy and will result in a smaller trade response, because the prices of non-traded goods must fall to hold GNP constant (Dornbusch pp. 868-70). Thus, the magnitude of trade elasticities depends on the prevailing macroeconomic policy.

In addition to all of the domestic variables, trade elasticities may depend on the reaction of foreign variables. Multi-country models have been constructed to capture the interdependence among national agricultural markets (Seeley; Anderson and Tyers; Tyers and Anderson). The domestic effects of monetary and fiscal policy may also depend on foreign repercussions. For example, the effect of fiscal policy on the exchange rate is sensitive to the presence or absence of foreign repercussion effects (Stevens et al).

EXCHANGE RATES AND TRADE

Several issues involving the effect of exchange rates on trade can be identified. An appropriate exchange rate measure must convert money rates into some kind of real exchange rate measure. The multitude of bilateral exchange rates must be converted into an effective exchange rate index. The separate effects of exchange rates on the volume of exports, foreign and domestic prices, and the value of exports must be determined.

Effective Exchange Rate Indices

If all bilateral exchange rates changed in the same proportions against the dollar, constructing an effective or average exchange rate index would be a simple task. Any bilateral rate would accurately represent the behavior of all the rates. There are some currency blocs that float jointly against the dollar. Examples are those EEC members of the European Monetary Union, and the former French colonies whose currencies are pegged to the French franc.

However, when all of the more than 150 currencies of the world are considered, it is common to observe changes in bilateral rates that vary substantially across countries. For example, the large depreciation of the U.S. dollar since March 1985 against the currencies of Japan and Western Europe has received much attention. However, the U.S. dollar changed very little against the Canadian dollar, and it appreciated against prominent trading partners in Asia and Latin America. From March 1985 to the end of 1986 the real U.S. dollar depreciated by 40 percent against an average of the Group of Ten countries (Belgium, United Kingdom, Canada, France, West Germany, Italy, Japan, Sweden, Netherlands, and the United States), but it appreciated by 3 percent against an average of eight prominent developing countries (Brazil, Mexico, Hong Kong, Malaysia, Philippines, Singapore, South Korea, Taiwan, see Pauls, p. 419). Thus, in constructing an effective exchange rate index the weights assigned to each country can be very important. This potential problem is common to the construction of all price indices.

There are many regularly published exchange rates indices available (see Dutton and Grennes 1985; Pauls), but most of the common ones (Federal Reserve Board, Special Drawing Rights, Morgan Guaranty Trust old version) include only the currencies of high income countries. Ott has shown that various indices obtained by varying weights within the Group of Ten countries behave nearly identically. He constructed alternative indices based on different weighting schemes for the ten countries in the Federal Reserve Board Index. Five indices were obtained using (1) trade shares for two periods, (2) capital flows for two periods, and (3) a naive index based on equal weights for all ten countries. For the period 1973-1981, the indices were nearly perfectly correlated, and out-of-sample forecasting errors using the indices in Belongia's agricultural export equation are indistinguishable from each other. Thus, country weights are not important for this set of currencies. However, all these indices assign zero weights to all the countries outside the Group of Ten, which includes many buyers of U.S. agricultural products and exporters of competing products.

The construction of exchange rate indices with broader country coverage has produced many alternative measures (Dutton and Grennes 1987; Pauls; Feldstein and Barchetta). Morgan Guaranty Trust has added a second index which includes more countries. Cox has produced an index that includes all the countries in the world (131) for which data are readily available. An alternative purely statistical approach to an exchange rate index has been attempted by Becketti and Hakkio.

The choice of country weights turns out to be important when weights are based on agricultural trade (Dutton and Grennes, 1987). The exchange rate index published in the Agricultural Outlook of the U.S. Department of Agriculture emphasizes low income countries. It is based on shares of the 37 largest national buyers of agricultural products from the United States. This index assigns small or zero weights to major competing exporters such as Canada, Australia, France, and Argentina. Conversely, when weights are based on global shares of agricultural trade, the relative importance of the competing exporters increases. Although bilateral and globally weighted indices are fairly highly correlated over long time periods, some large divergences have occurred since 1980 (Dutton and Grennes, 1987). Divergences between indices have generated different estimates of exchange rate elasticities and different out-of-sample forecast errors in agricultural export equations (Belongia).

In addition to altering the pattern of country weights there has been some experimentation with alternative base periods and chain link versus fixed weights indices. Trade weighted dollar indices seem less sensitive to these changes than the choice between bilateral and global weights. Whether an index uses arithmetic or geometric means is important when there are large differences in the movement of components. For example, nominal indices including countries with large inflation differentials will be quite different in arithmetic and geometric forms. Indices have also been constructed on the basis of weights for specific agricultural products such as wheat, corn, soybeans, and cotton (USDA; Dutton and Grennes, 1985; Henneberry, Drabenstott and Henneberry).

Nominal exchange rate indices must be adjusted for inflation to determine their effect on relative prices and the volume of trade. Otherwise, an index would give the misleading impression that countries with the highest inflation rates would gain a competitive advantage, since their currencies would depreciate the fastest. For example, the total agricultural trade-weighted dollar published in the Agricultural Outlook (June 1987, p. 55) had a value of 7783 in April 1987 (April 1971 = 100), indicating substantial nominal appreciation by the U.S. However, the real value of the same index was 81 in April 1987, indicating, moderate real depreciation of the dollar. The simplest way to adjust for inflation is to construct a real exchange rate index based on purchasing power parity. However, alternative concepts of the real exchange rate are common in the literature. Two frequent interpretations of the real exchange rate are (1) the relative price of traded and non-traded goods and (2) the terms of trade. Sebastian Edwards (1987) has identified at least five alternative measures which are imperfectly correlated and sometimes move in opposite directions.

The importance of adjusting for inflation depends on the inflation differentials for the countries being studied. Let the purchasing power parity real exchange rate be:

$$E_R = \frac{EP^*}{P}$$

where E is the nominal exchange rate and P and P* are the aggregate domestic and foreign price levels. Expressed as percentage changes, the real rate equals the nominal rate minus the inflation differential:

$$\frac{dE_R}{E_R} = \frac{dE}{E} - \left(\frac{dP}{P} - \frac{dP^*}{P^*}\right)$$

If domestic and foreign inflation rates are equal, the real rate equals the nominal rate. For indices (for example Federal Reserve Board) including only countries with similar inflation rates (Group of Ten or OECD), changes in nominal rates are similar to changes in real rates. However, broader indices that include both low and high inflation countries (for example, Mexico, Brazil, Argentina) show substantial differences between changes in nominal and real rates (U.S.D.A.; Cox).

If the real exchange rate is itself an endogenous variable, it may be difficult to isolate a relationship between the real exchange rate and other relative prices. The effect on the real rate and other relative prices will depend on the source of the shock to the system (Dutton, Stockman, Boughton et al). Sebastian Edwards constructed a general equilibrium model designed to show the effect of various disturbances on the relative price of non-traded goods. The disturbances considered were (1) an across-the-board tariff, (2) a change in the world price of importables, and (3) an exogenous capital flow. The impact on the real exchange rate and relative prices was different for each disturbance.

However it is measured, it has not been easy to explain the behavior of the real exchange rate. There is some evidence that it has been more volatile during the floating exchange rate period and it appears to follow an approximate random walk (Mussa, 1986; Hakkio, 1986). The question of whether fiscal deficits result in currency appreciation remains unresolved (Evans, Feldstein).

Exchange Rates and Prices

Currency depreciation raises relative prices of tradable goods expressed in domestic currency, and there has been both theoretical and empirical discussion of the magnitude of the price changes (Goldstein and Khan; Dornbusch, 1975). When all products are aggregated into a single group, the percentage increase of the price of the aggregate cannot exceed the percentage depreciation (Branson; Dornbusch, 1975). However, it is less clear that this restriction on average prices also holds for prices of individual products that are related through consumption and production (Dutton; Chambers and Just). The effect of depreciation on consumer prices has been called the pass-through effect of the depreciation. Exchange rate changes are completely passed through to consumers if a k percent depreciation results in a k percent reduction in foreign currency prices of exports and a k percent increase in domestic currency prices of imports.

The elasticities approach has been used to infer the effect of depreciation on domestic prices from the underlying domestic supply and demand elasticities (Goldstein and Khan, p. 1089; Branson, p. 21). The percentage change in the foreign currency price of exports (P_x^*) associated with a given change in the exchange rate (E) can be expressed as:

$$\frac{dP_x^*}{P_x^*} = \left(\frac{1}{1 - \frac{\eta_x}{\varepsilon_x}} \right) \frac{dE}{E}$$

where η_X and ε_X are the elasticities of demand and supply for exports. Unless the supply of exports is infinitely elastic, the depreciation will be incompletely passed through to foreign consumers. Because of the limit on export supply, the domestic currency price of exports will rise. As long as arbitrage equates the domestic currency price of exports and the domestic currency equivalent of the foreign currency price ($P_X = EP_X^*$), the percentage change in the domestic currency price can be inferred from the foreign currency price change:

$$\frac{dP_x}{P_x} = \frac{dE}{E} + \frac{dP_x^*}{P_x^*}$$

If the exporting country is a price-taker on the world market (ε_X is infinite), the domestic currency price rises by the full percentage rate of depreciation. In general, domestic and foreign currency prices are expected to move in opposite directions.

Some qualifications to the elasticities approach should be mentioned. First, the effect of an exchange rate change on relative prices depends on the prevailing monetary and fiscal policy (Dornbusch, 1975). Second, the real exchange rate may itself be an endogenous variable. In that case, the effect on prices may vary with the source of disturbances (Dutton; S. Edwards; Stockman; Boughton et al.). Third, it may be important to incorporate the interaction between current account and capital account transactions. For example, the asset market approach to the exchange rate (Frenkel and Mussa) focuses on stocks of money and financial assets rather than flows of exports and imports.

There is an alternative empirical literature in which export and import prices have been regressed on exchange rates (Goldstein and Khan, p. 1089; Pigott; Spitaeller; Dornbusch, 1985; Gardner). Empirical results generally support the proposition that the percentage change in prices does not exceed the percentage change in the exchange rate. A frequently cited exception is a study by Chambers and Just. The effects of exchange rates have also been simulated using multi-product models (Longmire and Morey; Krissoff and Morey; Johnson, Grennes, and Thursby).

A related empirical question is whether domestic prices respond in the same way to foreign prices as they do to exchange rate changes. The results are mixed (Goldstein and Khan, p. 1091; Wilson and Takacs; Richardson; Grennes and Lapp). One interpretation of the exchange rate in trade equations is that it represents the effect of omitted foreign prices. If both U.S. and the relevant foreign prices are expressed in dollars, the exchange rate need not appear separately. The use of both variables by Chambers and Just (plus the use of nominal SDRs to represent the exchange rate) may be responsible for their high exchange rate elasticities and low price elasticities. In empirical work in general it has been difficult to disentangle the effects of prices and exchange rates (C. Edwards).

Exchange Rates and the Quantity of Exports

Currency depreciation tends to increase the quantity of exports and decrease the quantity of imports, and the magnitude and timing of the trade response has been the subject of a large empirical literature (Goldstein and Khan; Magee; Stern, Francis, and Schumacher). For a given price in foreign currency, depreciation raises the price of exports in domestic currency. The increase in exports depends on the elasticity of export demand and the elasticity of export supply, both of which will be greater for more narrowly defined products. The long-run response is believed to be considerably greater than the short-run response.

Batten and Belongia estimated an equation for total U.S. agricultural exports as a function of foreign real income, real agricultural prices in the U.S., and the real exchange rate. Quarterly data from 1973-81 were used. In a separate paper (Belongia) five alternative exchange rate indices (Federal Reserve Board, MERM, Special Drawing Rights, Morgan Guaranty Trust, and USDA) were used to generate five sets of coefficients for the three explanatory variables. The ranges for the elasticity coefficients were (1) income +0.8 to +1.8, (2) price -0.4 to -1.0, (3) exchange rate -0.6 to -1.6. Different lags were used for each explanatory variable, and the lag lengths ranged from two to eight quarters.

At a lower level of product aggregation, Henneberry, Drabenstott, and Henneberry estimated the demand for U.S. wheat exports. In addition to income, U.S. price, and the exchange rate as explanatory variables they included a measure of foreign prices

(Australian wheat price) and the rest of the world's wheat production. The exchange rate index used weights based on wheat imports from the U.S. Quarterly data from the beginning of 1973 to 1986 (second quarter) were used. The estimated elasticities were: U.S. price -1.6, income +1.8, exchange rate -0.6, foreign price +1.6, and world wheat output -2.0.

Exchange Rate and the Value of Trade

A traditional question in the macroeconomic trade literature is whether currency depreciation would improve the balance of trade (Goldstein and Khan). According to the elasticities approach, the result depends on the elasticities of demand and supply of exports and imports. When values are expressed in foreign currency the value of imports must decline, but what happens to the value of exports depends on the relative importance of lower prices and greater quantities. When values are expressed in domestic currency, the value of exports necessarily rises (both price and quantity increase), but the value of imports may rise or fall depending on the relative importance of higher prices and lower import volume. If import prices rise faster than export prices, depreciation may be followed by a temporary deterioration of the trade balance. The so-called "J curve" is a popular representation of this transitory effect. The same general relationship should hold for particular exports or imports as for total trade, except that elasticities should be greater for narrowly defined products.

To the extent that the impact of an exchange rate change on trade depends on the underlying domestic elasticities, it is conditional on domestic agricultural policy. The response of exports to currency depreciation will be smaller if there are production or acreage controls. The effect of binding price supports can be represented by either altering the elasticity of export supply or the elasticity of inventory demand. The dynamic path of adjustment may also be affected by price expectations. For example, lower price supports in the United States were announced in December 1985, but they did not take effect for grain until June-September of 1986. Thus, for the first six months following the announcement prices were expected to decline. U.S. grain exports declined, and for a three month period in 1986 there was the first agricultural trade deficit in decades (Grennes).

THEORETICAL ISSUES INVOLVING THE EXCHANGE RATE

One cannot interpret a statistical relationship between exchange rates and other variables without the guidance of an underlying economic model. Theoretical considerations might also help in choosing the most useful empirical representation of a theoretical variable. Choosing one among many alternative exchange rate measures is particularly relevant to the study of agricultural trade.

The Exchange Rate in a Simple Three-Good Model

To illustrate the effect of an exchange rate, we employ a simple three good, two country model. Foreign demand for U.S. agricultural exports is assumed to take the following form:

$$X = \tilde{f}\left(P_A^*, P_O^*, P_N^*, Y^*\right) \qquad (1)$$

where P_A^* = price of U.S. agricultural export good,

P_O^* = price of other tradeable good,

P_N^* = price of foreign nontradeable good, and

Y^* = income of foreign country.

Starred variables are expressed in foreign currency. This excess demand equation, if properly specified, should be homogeneous of degree zero in prices and income, even in aggregate form. This characteristic allows multiplication of all the arguments by a common factor. For example, we can multiply all of them by E, the nominal exchange rate expressed in dollars per unit of foreign currency. The result is the demand function with the prices expressed in dollars:

$$X = \tilde{f}\left(P_A, P_O, P_F, Y\right) \tag{2}$$

It is notable that neither of these equations contains the exchange rate. In the transformation of prices and income from foreign currency to dollars the exchange rate disappears because of zero degree homogeneity of demand. In fact, an exchange rate in either of these equations would be a specification test of sorts. A significant coefficient would indicate the strong possibility of misspecification, probably from left-out variables correlated with the included regressors.

We can use zero degree homogeneity again in (2) to deflate by a numeraire price. Let the price level in the U.S., P, be the numeraire. Dividing each argument by P gives:

$$X = \tilde{f}\left(\frac{P_A}{P}, \frac{P_O}{P}, \frac{P_N}{P}, \frac{Y}{P}\right) = f\left(p_A, p_O, p_N, y\right) \tag{3}$$

where lower case p's and y's indicate deflated values. These price and income levels can then be considered as real values.

The exchange rate does not appear explicitly. The zero degree homogeneity property of demand implies that the nominal exchange rate will not play a role in such an equation if prices and income are all expressed in the same currency.

Many analysts have entered the <u>real</u> exchange rate in equations like (3). The real exchange rate (E_T) is frequently defined as follows:

$$E_R = \frac{EP_R^*}{P}$$

where P_R^* and P are the nominal price levels abroad and at home, measured in foreign and home currency. This can also be written:

$$E_R = \frac{P_R}{P} = p_R \tag{4}$$

where P_R and P are in nominal dollars, and p_R is in dollars deflated by the U.S. price level. Define a fixed weight price index for the rest of the world:

$$P_R^* = \alpha_A p_A^* + \alpha_O p_O^* + \alpha_N p_N^*$$

where α_i is the expenditure share of good i. Convert to dollars and deflate by the U.S. price level to get:

$$p_R = \alpha_A p_A + \alpha_O p_O + \alpha_N p_N. \tag{5}$$

Using (4) results in an expression for P_N:

$$p_N = \frac{E_R - \alpha_A p_A - \alpha_O p_O}{\alpha_N}.$$

We can substitute this into (3) to obtain:

$$X = f\left(p_A, p_O, \left(\frac{E_R - \alpha_A p_A - \alpha_O p_O}{\alpha_N}\right), y\right). \tag{6}$$

This foreign demand can be specified as a function of the real exchange rate through this substitution process. This specification does, however, introduce some complexity into the implied coefficients, so that:

$$X_E = \frac{f_N}{\alpha_N}$$

$$X_A = f_A - \alpha_A X_E$$

$$X_O = f_O - \alpha_O X_E$$

where X_i is the partial derivative of excess demand with respect to price i, and f_i is the partial of f with respect to price i.

The coefficient of the real exchange rate reflects the effect of the foreign non-traded good price. That effect is multiplied by the inverse of the share of good N in the foreign economy. The presence of the exchange rate also biases the other two price effects. The own price effect of the agricultural good lowered if X_E is positive as expected (or raised if X_E is negative). If α_N is large, then the biases in X_A and X_O will be small. Also in that case the effect of E_r will be very close to f_N. The real exchange rate will serve to represent the price level of "the" alternative good in the foreign economy.

We can see from (5) and (6) that other specifications of the estimated equation are possible. We could have substituted for p_O rather than p_N, for example. In such a case the measured real exchange rate effect would have been changed to $E_r = p_O/\alpha_O$. Also, however, the set of other prices in the equation, p_A and p_N, would be different from (6). The actual coefficients estimated would depend on what other prices were included with the real exchange rate in the estimated equation. The specification chosen would generally result from convenience of estimation.

One specification which will prove to be useful is to enter real income in terms of foreign currency rather than dollars. (The dollar denominated real income variable turns out to be highly collinear with the exchange rate.) In such a case, the interpretation of the exchange rate coefficient is still different. Equation (6) takes the form:

$$X = f\left(p_A, p_O, \left(\frac{E_R - \alpha_A p_A - \alpha_O p_O}{\alpha_N} \right), E_R y^* \right) \tag{7}$$

where y^* is real income in foreign currency. The partial derivative of X with respect to E_R is then:

$$E_R = \frac{f_N}{\alpha_N} + f_y y^*.$$

If the specification above is correct, the real exchange rate coefficient should be picking up an income component as well as the foreign good price effect.

Relative to this model the real exchange rate is simply a function of other prices. It is not shocks to the real exchange rate which affect demand, but shocks to the individual prices (Dutton). In interpreting the effects of the exchange rate in an equation such as (6) or (7), one must note what other prices are in the equation and remember that the exchange rate effect is being observed under an implicit assumption that the other prices in the equation are held constant.

For the above specification of the export equation the relevant exchange rate is between the dollar and the currency of the purchasing country. It also indicates that the exchange rate should be real. Using a nominal rate in such an equation is most likely a mismeasurement and may well lead to bias in the coefficients. In empirical work the foreign country is actually likely to be an aggregation of purchasers of U.S. goods. As indicated in Dutton and Grennes (1985), the exchange rate used is generally an index of exchange rates with a set of other countries. There are many possible mathematical forms for such an index. Unfortunately, neither the theory underlying the export equation above, nor other theory, reveals grounds for choosing one form over another.

The Exchange Rate in a Four-Good Model

A richer model would be one with an additional "competitor" good, available from a third country. This good is a fairly close substitute for the U.S. agricultural good, perhaps a differentiated version of the same good. The function is:

$$X = \tilde{f}\left(P^*_A, P^*_O, P^*_N, P^*_C, Y^*\right) \tag{8}$$

where P^*_C is the price of the competing good in terms of the currency of the customer country. As before, zero degree homogeneity can be cited to justify conversion of all financial variables to dollars. Alternatively, one might want other equation forms. For example, the competitor price in dollars could be replaced by (P°_C * E°), where P°_C is the competitor price in competitor country currency and E° is the dollar-competitor currency exchange rate, expressed in dollars per unit of competitor currency. This form would justify including an exchange rate in the equation, even though all variables are nominal. To be properly specified the equation should also contain P°_C. If that term were omitted, the exchange rate would pick up its effects. Its coefficient and others would be biased if P°_C were correlated with the other variables. It is wise to remember that the measured exchange rate effect would represent the influence of a left out price variable. This is one sense in which a significant exchange rate coefficient in an equation with nominal values indicates a likely misspecification.

Proceeding further with this form, we can deflate most variables by p as before. Adding the price level of other goods (empirically, the price level in general) of the competitor country, gives:

$$X = f\left(p_A, p_O, \left(\frac{E_R - \alpha_A p_A - \alpha_O p_O - \alpha_C p_C}{\alpha_N}\right), E^\circ_R P^\circ_C, y\right) \tag{9}$$

where P°_{C} is the real price of the competitor good, deflated by the price level in the competitor country, and

$$E^{\circ}_{R} = \frac{P^{\circ}_{C}E^{\circ}}{P}$$

is the real exchange rate between the dollar and the competitor currency. This equation contains two separate exchange rates as arguments, the exchange rate with the customer country and the exchange rate with the competitor country.

It is quite likely that these two exchange rates are highly correlated. Certainly they both contain effects of whatever influences originate in the U.S. If the correct model is (9) but one of the two exchange rates is left out, measured effects of the other are likely to be biased. The meaning of a coefficient in such a case would not be clear. Also, a wide variety of effective exchange rate measures might prove statistically significant, but with varying magnitude of effect.

These simple formulations illustrate the point that empirical relationships between the real exchange rate, relative prices, and the volume of trade may be difficult to interpret. Introducing money and net foreign assets would make the relationships involving exchange rates more complex. Changes in money supply, money demand, and capital flows (S. Edwards; Boughton et al) would be additional sources of disturbances. There is evidence that changes in the money supply affect the exchange rate before the price level changes (Mussa). Exchange rates behave like other asset prices in efficient markets. Aggregate price levels behave sluggishly and exhibit serial correlation. Thus, the real exchange rate may change in absence of shocks to demands or supplies of individual products. It has been observed that the purchasing power parity real exchange rate has been more variable since the advent of floating exchange rates than during the earlier Bretton Woods period. However, it remains to be seen whether increased volatility of the real rate is due to monetary disturbances or real disturbances.

ESTIMATIONS

Equations for foreign demand for aggregate agricultural exports, as well as equations for three export crops, are presented in Tables 4.6-4.9. All financial variables are in real (deflated) terms. Quantity and value variables are divided by an index of population of chief purchasers of U.S. agricultural exports; this division should eliminate or reduce a major potential source of heteroskedasticity. Tables 4.6 and 4.7 contain the estimation results for aggregate exports. Regressors include an index of prices of U.S. agricultural exports, an index of world agricultural prices, an index of real GDP per capita of foreign purchasers, an index of the real exchange rate between the dollar and currencies of purchasing countries, quarterly dummies, and a time trend term. All variables but time and quarterly dummies are expressed in logs. For the four chief regressors, a polynominal distributed lag specification of order two is used. Lag length has been arbitrarily set at seven quarters. This second order PDL reduces degrees of freedom loss while allowing for either of the two most economically plausible lag structures, i.e., monotonically declining coefficients or initially increasing followed by monotonically declining coefficients.

Tables 4.8 and 4.9 present estimation results for three individual crops. Quantity of exports per capita is regressed on U.S. prices and representative foreign prices, an index of real GDP per capita and a real exchange rate index expressed in dollars per unit of foreign currency. Real GDP and population indexes, as well as real exchange rates, are derived for each crop using data from principal purchasers of that crop. Quarterly dummies and a time trend are included for each crop. The same PDL lag specification is used as for the total agriculture equations.

A problem of data arose from the lack of quarterly versions of some series. Ginsburgh (1973) describes ways of converting annual data into quarterly. His technique is used here to derive a quarterly version of the GDP index. In addition the foreign population series was derived with the method of Boot et al. described in Ginsburgh.

Several econometric problems are potentially present. One is autoregressive errors, which can cause bias in standard errors of the estimated parameters. Another potential problem is endogeneity of regressors. In the present instance, we suspect the domestic price term of being potentially endogenous.

Using standard SAS procedures, we estimate each equation in both an "OLS" and a "2SLS" version. In each case we first estimate an autoregressive structure for the errors, with AR terms up to seven, and drop the higher order terms where they are not significant. For the "2SLS" version, we use instruments for the own price terms in this process. We then re-estimate using transformed versions of the data. The reported equations are the results of this stage.[1]

Another potential econometric problem is the use in the total agricultural export equation of a unit value index along with a quantity index derived from it. This method of derivation could potentially bias the own price results upward (make the own price effect more negative) in that equation.

Several conclusions emerge from the results. First, the real exchange rate terms do not perform as predicted. We expected the dollar value of foreign currency to evidence a positive coefficient for two reasons. The positive cross price effect from foreign non-agricultural goods should be represented by the real exchange rate. Also, since the real income variable is in foreign currency terms, the exchange rate coefficient picks up a positive income effect as well. However, in our equations, the exchange rate effect, as measured by coefficient sums, is uniformly negative. In no case was the effect statistically significant at the 5 percent level, and only once was it so at the 10 percent level.

When a variable does not perform as theory predicts, one explanation is that the theory is wrong; we are certainly not convinced by the evidence at hand that economic theory is predicting the wrong sign for the exchange rate effect. Another interpretation is that there are important left-out regressors which are correlated with the exchange rate. One can argue that the real exchange rate changes only when major shocks occur in the economy. The rate is altered only by important intersectoral demand or supply shifts, caused by substantial shocks to monetary or fiscal policy, technological change, or similar events. Those shocks in all likelihood have strong effects on agricultural exports, as well as on the real exchange rate. Our simple regressions cannot control for the wide variety of shocks which may have occurred. The exchange rate, highly correlated with those shocks, exhibits strongly biased and unstable effects.

Turning to the remaining variables, we can note that own price, cross price, and income effects are for the most part consistent with our expectations, at least in sign. The most notable exception is the negative (but not significant) income effect in the wheat equations. A second exception is the approximately zero own price effect in the "OLS" corn equation.

The equations indicate relatively high own price elasticities, with the exception of corn. Cross price effects from foreign prices of the same products tend to be of the same order of magnitude as own price effects. Since the commodities represented are very close substitutes, the own and cross price variables are highly collinear. Consequently, we can expect the effects to be difficult to estimate with precision.

Similarly, per capita real income and time exhibit substantial collinearity. For the most part, the equations indicate a negative effect of time and a strong positive income effect. Wheat is the exception; the signs of the two are reversed in that equation. One general interpretation of these results is of a secular decline in foreign demand caused by technological change, accompanied by counteracting results of income growth. An alternative explanation is that collinearity between time and income has resulted in "too large" an estimate for the effect of each. We have not been able to distinguish between these two hypotheses.

If our equations exhibit some anomolous results, it is perhaps not completely surprising. Above we have mentioned several difficulties in estimating export demand equations. A major one is aggregation. Probably the cross-country aggregation of data used for our estimations is more problematical than cross-firm or cross-individual aggregation within a single country would be. As we point out earlier, a given movement of the U.S. price may mean quite different things in different foreign countries, depending on such factors as differences in policy interventions.

Policy interventions themselves are another explanation. It is well known that governments have often systematically sought to reduce the domestic effects of international price movements.

A third problem is the shortness of our data series. To maximize homogeneity of economic structure, and because most exchange rate changes have occurred in the 1970s and 1980s, we have used data from 1969 on. The data are quarterly; however, given the annual periodicity of agricultural markets, we may be observing only a much smaller number of independent events. In other words, our quarterly data may not be yielding much more information than annual data would.

A fourth problem is the nature of the available data. Every empirical economist complains about data. In this case, however, the data may be worse than usual. They, of necessity, come from a wide variety of sources. Several variables are constructed as indexes combining data from many countries. And in some cases quarterly data had to be constructed from national data (though only for real GDP and population, variables likely to be fairly smooth in their movements).

SUMMARY

There is no single trade elasticity that applies to all products in all circumstances. The response of trade to a price change depends on both the elasticity of export demand and the elasticity of export supply, and they are conditional on many things. Instead of a single elasticity, there is a matrix of possible trade responses, each of which depends on the conditions prevailing at the time of the price change. The wide range of elasticities reported in the literature is partly attributable to differences in assumptions about the underlying conditions. Thus, it is meaningless to ask whether the demand for agricultural exports from the United States is elastic or not without specifying the conditions under which the response will occur.

The response of trade depends on the length of the time period being considered and the completeness of price transmission to various foreign countries. It depends on whether the good considered is a narrowly defined product whose price changes relative to a wide array of other products or a broad aggregate with limited scope for relative price change. The outcome also depends on the completeness of price transmission to domestic producers and consumers, the severity of restrictions on domestic production and input use, and the nature of government inventory policy.

An important issue in empirical work is interpretation of exchange rate effects. The nominal exchange rate is not a goods price which would normally appear in foreign excess demand. The real exchange rate, on the other hand, may represent the price of foreign goods, as well as other effects (depending on the other variables in the equation). It is important to interpret exchange rate coefficients in the light of the model specification.

Empirical results confirm the importance of relative prices and income on trade, but it is difficult to separate the effects of prices and exchange rates. The importance of time lags is also confirmed.

Isolating the impact of exchange rates is complicated by the problem of choosing an exchange rate index. The possible endogeneity of the exchange rate may make it difficult to estimate a consistent relationship between exchange rates and trade. In that case, the effect of the exchange rate may vary according to the source of the disturbance.

Ideally, macroeconomic shocks should be controlled for, along with the exchange rate in an export demand equation. A significant avenue for future research is to find ways to incorporate such disturbances in a model so as to obtain a better understanding of exchange rate effects on agricultural exports.

Table 4.1. Effects on Prices and Quantities (Percent Change) of a 10 Percent Real Appreciation

Price or quantity	Zero cross-commodity effects	Consensus cross-commodity effects
Wheat price	-10	-7
Corn price	-7	-7
Soybean price	-5	-6
Wheat exports	-7	-3
Corn exports	-13	-2
Soybean exports	-9	-3

Consensus Elasticities of Demand

Elasticity with respect to price of:	Wheat	Corn	Soybeans
Wheat	-0.20	0.05	0.05
Corn	0.05	-0.40	0.10
Soybeans	0.05	0.10	-0.40

Consensus Elasticities of Supply

Elasticity with respect to price of:	Wheat	Corn	Soybeans
Wheat	0.40	-0.15	-0.05
Corn	-0.15	0.40	0.30
Soybeans	-0.05	-0.30	0.40

Source: Longmire and Morey, p. 30-31.

Table 4.2. A Comparison of Price Changes and Selected Impact Elasticities Resulting from a 5 Percent Supply Reduction for the United States from Six International Trade Models for Wheat, Coarse Grains and Soybeans (One Year Response)

	WTM(a)	FAPRI(b)	MSU(b)	GOL(b)	IIASA(b)	WB(b)
Wheat						
Price change (percent)	5.5	17.1	1.5	4.6	3.6	1.5
Export demand elasticity	-1.0	-0.3	1.0	-1.7	-1.7	-0.4
Stock demand elasticity	N.I.	-0.3	-6.5	N.I.	-1.0	-5.9
Domestic demand elasticity(c)	N.I.	-0.4	N.R.	-0.5	-0.3	0.1
Total demand elasticity	N.I.	-0.2	N.R.	-1.1	N.R.	-2.0
Coarse Grain(d)						
Price change (percent)	N.I.	12.9	7.9	7.8	9.2	2.3
Export demand elasticity	N.I.	-1.0	-0.2	-1.1	-1.2	-0.5
Stock demand elasticity	N.I.	-0.2	-1.3	N.I.	-1.8	-4.7
Domestic demand elasticity	N.I.	-0.2	N.R.	-0.4	Neg.	-0.1
Total demand elasticity	N.I.	-0.3	N.R.	-0.6	N.R.	-1.5
Soybeans(e)						
Price change (percent)	N.I.	12.9	20.8	20.2	13.4	4.4
Export demand elasticity	N.I.	-0.4	-0.1	-0.5	-0.5	-0.1
Stock demand elasticity	N.I.	-0.3	N.R.	N.I.	-0.4	-8.5
Domestic demand elasticity	N.I.	-0.3	N.R.	-0.1	-0.2	-0.5
Total demand elasticity	N.I.	-0.3	N.R.	-0.3	N.R.	-1.1

WTM = World Wheat Trade Model (Dixit and Sharples)
FAPRI = Food and Agricultural Policy Research Institute (Meyers, Devadoss and Helmar)
MSU = Michigan State University (Shagram)
GOL = Grain, Oilseeds, Livestock Model (Liu and Roningen)
IIASA = International Institute for Applied Systems Analysis (Frohberg)
WB = World Bank (Mitchell)
N.I. = not included
N.R. = not reported

(a) Partial elasticities, no cross-commodity effects allowed.
(b) Total elasticities, cross-commodity effects included.
(c) For FAPRI feed wheat only.
(d) For GOL and FAPRI corn only.
(e) For the IIASA model the results for protein feeds are reported. The FAPRI, WB, and GOL models also include endogenous soybean oil and meal sectors while the MSU model includes an endogenous soybean meal sector.

Source: K. Meilke. "A Comparison of the Simulation Results from Six International Trade Models," February 1987.

Table 4.3. Elasticities of Export Demand for Wheat Facing Various Countries, 1985-86 (Quantities in Million Metric Tons)

Country	Production	Exports	Demand rest of world	Supply rest of world	η_x^{*}	η_x^{**}
	Q^s	X	Q_R^d	Q_R^s	η_E^{*}	η_E^{**}
United States	66	25	458	433	-7.2	-3.7
Argentina	9	4	494	490	-49.2	-24.7
Australia	16	16	499	483	-12.3	-6.2
Canada	24	18	492	474	-10.7	-5.4
EEC-12	72	16	443	427	-10.9	-5.5
Five Exporters	187	78	390	312	-1.8	-1.0
World	499	96	0	0	-0.2	--

$$\eta_X^{*} = \frac{Q_R^d}{X}\eta - \frac{Q_R^s}{X}\varepsilon$$

**Short-run when $\varepsilon = 0$.

The same values ($\eta = -.2$, $\varepsilon = +.2$) are assumed for all countries. Since production is assumed equal to consumption, inventory changes are not taken into account.

Source: U.S. Wheatletter Associates, February 1987.

Table 4.4. Simulated Effect of 10 Percent Real Depreciation of Dollar (Percent Changes)

	Against imports only	Against exports only	Against all currencies
Wheat price	6.5	2.1	8.9
Corn price	5.5	2.3	8.0
Soybean price	5.2	1.7	7.1
Wheat exports	5.2	1.5	7.7
Corn exports	8.1	3.6	12.2
Soybean exports	2.9	0.8	3.7

Source: Krissoff and Morey, p. 8.

Table 4.5 Effects of Incomplete Price Transmission Following 10 Percent Real Depreciation Against All Currencies

	Complete transmission	Incomplete* transmission
Wheat price	8.9	7.2
Corn price	8.0	7.4
Soybean price	7.1	6.7
Wheat exports	7.1	5.1
Corn exports	11.8	11.2
Soybean exports	3.7	3.7

*Price transmission elasticities: Wheat = .5
Corn = .8
Soybeans = .9

Source: Krissoff and Morey, p. 10-11.

Table 4.6. Estimation Results for Total Agricultural Exports, 1969-1986 Second Degree Polynomial Distributed Lag Estimation (Absolute Values of t-Statistics in Parentheses)

		"OLS"		"2SLS"	
U.S. Price		-0.039	(0.13)	-0.002	(0.01)
	Lag 1	-0.166	(1.08)	-0.232	(1.23)
	Lag 2	-0.258	(2.62)	-0.381	(2.99)
	Lag 3	-0.316	(2.83)	-0.450	(2.83)
	Lag 4	-0.339	(3.12)	-0.438	(2.87)
	Lag 5	-0.328	(4.03)	-0.346	(3.36)
	Lag 6	-0.282	(2.37)	-0.173	(1.06)
	Lag 7	-0.202	(0.76)	0.080	(0.21)
R.O.W. Price		0.037	(0.11)	0.072	(0.18)
	Lag 1	0.138	(0.75)	0.19	(0.83)
	Lag 2	0.226	(1.88)	0.283	(1.78)
	Lag 3	0.301	(2.59)	0.353	(2.25)
	Lag 4	0.363	(3.25)	0.397	(2.65)
	Lag 5	0.413	(4.05)	0.418	(3.10)
	Lag 6	0.449	(2.96)	0.414	(2.08)
	Lag 7	0.473	(1.64)	0.385	(1.02)
R.O.W. Income		3.316	(2.43)	2.897	(1.87)
	Lag 1	1.572	(2.38)	1.421	(1.77)
	Lag 2	0.271	(0.49)	0.341	(0.45)
	Lag 3	-0.586	(0.81)	-0.342	(0.36)
	Lag 4	-0.999	(1.34)	-0.628	(0.65)
	Lag 5	-0.969	(1.64)	-0.519	(0.64)
	Lag 6	-0.495	(0.97)	-0.012	(0.02)
	Lag 7	0.422	(0.40)	0.891	(0.69)
Exchange Rate		0.059	(0.14)	-0.284	(0.60)
	Lag 1	-0.022	(0.15)	-0.095	(0.52)
	Lag 2	-0.088	(0.43)	0.011	(0.05)
	Lag 3	-0.141	(0.49)	0.035	(0.11)
	Lag 4	-0.178	(0.67)	-0.023	(0.08)
	Lag 5	-0.202	(1.37)	-0.164	(0.92)
	Lag 6	-0.211	(1.14)	-0.387	(1.70)
	Lag 7	-0.206	(0.38)	-0.692	(1.12)
Time		-0.004	(0.22)	-0.013	(0.46)
Intercept		-4.261	(0.24)	-8.608	(0.40)
Quarter 1		-0.014	(0.58)	-0.091	(0.74)
Quarter 2		-0.104	(4.15)	-0.108	(4.09)
Quarter 3		-0.193	(8.32)	-0.20	(7.75)
AR1(a)		0.329		0.395	
AR2		-0.157		-0.098	
AR3		0.202		-0.277	
AR4		-0.277		-0.176	
$\bar{R}^2$		0.92		0.89	
DW		2.12		2.19	

(a) Autoregressive parameters estimated in earlier stage regression.

Table 4.7. Total Effects of Lagged Variables in Equations for Total Agricultural Exports

		P Exogenous "OLS"	P Endogenous "2SLS"
U.S. Price:	Sum	-1.931	-1.943
	F	10.2359	5.8419
	Prob.	0.0027	0.0202
R.O.W. Price:	Sum	2.400	2.512
	F	7.0023	4.4977
	Prob.	0.0115	0.0400
R. O. W. Income:	Sum	2.532	4.051
	F	0.6790	0.8640
	Prob.	0.4147	0.3581
Exchange Rate:	Sum	-0.990	-1.600
	F	1.5331	2.3051
	Prob.	0.2227	0.1366

Table 4.8. Estimation Results for Wheat, Corn, and Soybean Exports, 1969-1986 Second Degree Polynomial Distributed Lag Estimation (Absolute Values of t-Statistics in Parentheses)

	Wheat		Corn		Soybeans	
	"OLS"	"2SLS"	"OLS"	"2SLS"	"OLS"	"2SLS"
U.S. Price	0.160	-0.302	0.046	0.295	-0.425	-0.225
	(0.29)	(0.43)	(0.28)	(1.48)	(2.88)	(0.94)
Lag 1	-0.409	-0.673	-0.50	0.065	-0.373	-0.249
	(1.33)	(1.81)	(0.54)	(0.65)	(3.51)	(1.36)
Lag 2	-0.765	-0.851	-0.105	-0.082	-0.324	-0.252
	(2.96)	(2.59)	(1.33)	(0.85)	(3.13)	(1.48)
Lag 3	-0.908	-0.837	-0.119	-0.148	-0.276	-0.232
	(3.01)	(2.13)	(1.36)	(1.25)	(2.53)	(1.42)
Lag 4	-0.838	-0.629	-0.090	-0.132	-0.230	-0.189
	(2.66)	(1.60)	(1.06)	(1.18)	(2.17)	(1.33)
Lag 5	-0.556	-0.229	-0.021	-0.034	-0.186	-0.124
	(1.92)	(0.69)	(0.27)	(0.40)	(1.88)	(1.07)
Lag 6	0.061	0.363	0.090	0.146	-0.144	-0.037
	(0.19)	(0.93)	(0.93)	(1.30)	(1.27)	(0.26)
Lag 7	0.648	1.149	0.243	0.408	-0.104	0.072
	(1.29)	(1.58)	(1.38)	(1.72)	(0.60)	(0.29)
R.O.W. Price	0.422	0.912	0.338	0.128	0.16	0.005
	(0.92)	(1.59)	(1.69)	(0.54)	(1.13)	(0.02)
Lag 1	0.480	0.716	0.191	0.070	0.229	0.159
	(2.08)	(2.57)	(1.85)	(0.62)	(2.15)	(1.04)
Lag 2	0.491	0.523	0.079	0.022	0.266	0.251
	(2.38)	(2.00)	(1.01)	(0.26)	(2.68)	(2.08)
Lag 3	0.457	0.332	0.001	-0.013	0.277	0.280
	(1.74)	(0.98)	(0.01)	(0.12)	(2.66)	(2.26)
Lag 4	0.376	0.144	-0.041	-0.038	0.261	0.246
	(1.37)	(0.42)	(0.45)	(0.34)	(2.50)	(1.94)
Lag 5	0.249	-0.042	-0.049	-0.052	0.220	0.150
	(1.50)	(0.15)	(0.65)	(0.60)	(2.21)	(1.25)
Lag 6	0.076	-0.225	-0.022	-0.054	0.151	-0.009
	(0.32)	(0.79)	(0.24)	(0.54)	(1.48)	(0.07)
Lag 7	-0.144	-0.406	0.040	-0.046	0.057	-0.230
	(0.35)	(0.72)	(0.21)	(0.21)	(0.42)	(1.24)
R.O.W. Income	0.331	-0.187	5.736	4.899	2.599	2.769
	(0.17)	(0.08)	(4.15)	(2.90)	(1.54)	(1.37)
Lag 1	0.053	-0.415	2.790	2.710	1.786	1.616
	(0.06)	(0.44)	(5.22)	(4.56)	(2.59)	(2.05)
Lag 2	-0.155	-0.540	0.639	1.057	1.137	0.776
	(0.20)	(0.57)	(1.39)	(1.84)	(1.83)	(1.01)
Lag 3	-0.294	-0.562	-0.716	-0.060	0.653	0.250
	(0.27)	(0.42)	(1.02)	(0.07)	(0.73)	(0.22)

Table 4.8. Continued.

	Wheat		Corn		Soybeans	
	"OLS"	"2SLS"	"OLS"	"2SLS"	"OLS"	"2SLS"
Lag 4	-0.364	-0.481	-1.277	-0.641	0.332	0.036
	(0.33)	(0.35)	(1.81)	(0.71)	(0.36)	(0.03)
Lag 5	-0.364	-0.298	-1.042	-0.687	0.176	0.136
	(0.41)	(0.29)	(2.33)	(1.25)	(0.27)	(0.18)
Lag 6	-0.295	-0.011	-0.013	-0.198	0.184	0.549
	(0.31)	(0.01)	(0.03)	(0.46)	(0.28)	(0.68)
Lag 7	-0.156	0.378	1.812	0.827	0.357	1.275
	(0.08)	(0.16)	(1.50)	(0.54)	(0.23)	(0.63)
Exchange Rate	0.218	-1.499	-0.400	-0.740	-1.119	-1.368
	(0.17)	(0.92)	(0.62)	(1.07)	(1.59)	(1.65)
Lag 1	-0.736	-2.175	-0.276	-0.751	-0.777	-0.705
	(0.74)	(1.73)	(0.65)	(1.48)	(1.65)	(1.34)
Lag 2	-1.328	-2.433	-0.187	-0.731	-0.547	-0.287
	(1.29)	(1.79)	(0.43)	(1.38)	(1.09)	(0.48)
Lag 3	-1.557	-2.271	-0.131	-0.679	-0.428	-0.115
	(1.40)	(1.57)	(0.26)	(1.15)	(0.73)	(0.15)
Lag 4	-1.424	-1.691	-0.109	-0.594	-0.421	-0.188
	(1.29)	(1.27)	(0.19)	(0.95)	(0.68)	(0.22)
Lag 5	-0.927	-0.692	-0.120	-0.478	-0.525	-0.506
	(0.92)	(0.62)	(0.20)	(0.74)	(0.86)	(0.58)
Lag 6	-0.067	0.726	-0.165	-0.329	-0.741	-1.070
	(0.07)	(0.56)	(0.24)	(0.43)	(1.12)	(1.14)
Lag 7	1.155	2.563	-0.244	-0.148	-1.069	-1.880
	(0.80)	(1.14)	(0.27)	(0.14)	(1.17)	(1.58)
Time	0.028	0.021	-0.017	-0.019	-0.048	-0.047
	(1.48)	(1.86)	(2.54)	(2.80)	(3.33)	(2.69)
Intercept	21.521	33.071	-23.806	-31.287	-32.438	-9.678
	(1.21)	(2.94)	(2.20)	(3.49)	(2.54)	(0.81)
Quarter 1	-0.085	-0.081	0.017	-0.041	-0.052	-0.044
	(1.78)	(1.51)	(0.24)	(0.59)	(0.52)	(0.55)
Quarter 2	0.001	-0.011	0.019	-0.029	-0.307	-0.248
	(0.02)	(0.16)	(0.23)	(0.32)	(3.22)	(2.43)
Quarter 3	0.160	0.190	-0.008	-0.096	-0.764	-0.744
	(3.35)	(3.21)	(0.10)	(1.28)	(7.73)	(9.05)
AR1(a)	0.332	0.451	-0.024	-0.023		
AR2	-0.285		-0.387	-0.369		
AR3				-0.090		
AR4				-0.010		
AR5				-0.286		
R	0.47	0.49	0.91	0.93	0.77	0.74
DW	1.85	2.15	2.40	2.32	2.18	1.94

(a) Autoregressive parameters estimated in earlier stage regression.

Table 4.9. Total Effects of Lagged Variables in Equations for Agricultural Crop Exports

		Wheat		Corn		Soybeans	
		"OLS"	"2SLS"	"OLS"	"2SLS"	"OLS"	"2SLS"
U.S. Price:	Sum	-2.7290	-2.0420	-0.0060	-0.5183	-2.0610	-1.2360
	F	2.9608	1.5980	0.0001	0.6350	9.5733	2.0696
	Prob.	0.0925	0.2132	0.9915	0.4302	0.0034	0.1573
R.O.W. Price:	Sum	2.4070	1.9538	0.5355	0.0161	1.6256	0.8523
	F	4.2061	2.8279	0.8102	0.0006	6.6589	1.2007
	Prob.	0.0464	0.1001	0.3732	0.9803	0.0133	0.2791
R. O. W. Income:	Sum	-1.2440	-2.1160	7.9287	7.9078	7.2249	7.4066
	F	0.4609	1.2416	24.8552	22.7864	11.9789	9.7885
	Prob.	0.5008	0.2715	0.0001	0.0001	0.0012	0.0031
Exchange Rate:	Sum	-4.6660	-7.4736	-1.6323	-4.3500	-5.6285	-6.1188
	F	0.6471	1.6252	0.2566	1.5130	3.1289	2.0562
	Prob.	0.4256	0.2094	0.6151	0.2260	0.0838	0.1587

VARIBLES USED IN ESTIMATION PROCEDURES

Total agricultural exports is a quantity index computed by the U.S. Bureau of the Census. The index is computed by dividing value by a unit value index. For use in estimation, the index has been divided by a population index for the main importers of U.S. agricultural production.

The agricultural export price is a unit value index provided by the U.S. Bureau of the Census. We have deflated it by the U.S. CPI.

The foreign agricultural price is from an index of food prices published in the U.N. Monthly Bulletin of Statistics. The U.N. index is in dollar form. For our use it is deflated by the U.S. CPI.

The real exchange rate variable is a geometric index derived using 1976-78 agricultural export weights from the U.S. Department of Agriculture. The inflation adjustment is done using CPI's from the relevant countries. Exchange rate and price index data are taken from the I.M.F. International Financial Statistics tape. Data for Taiwan come from a publication of the Taiwan central bank. A separate index is computed for each crop as well, using the countries which are primary purchases of that crop.

Foreign real GDP is computed from real GDP data in the I.M.F. International Financial Statistics tape. Quarterly data are used where available; annual data are used where necessary. The technique of Ginsburgh is used to interpolate. That technique makes use of the quarterly data which are available to derive a quarterly form for the other data. The resulting real GDP index is in terms of domestic currencies of the countries involved. We divide the index by an index of population of the countries involved to convert to per capita terms. A similar index is used for each crop.

Wheat, corn and soybean quantities are taken from various issues of the Commodity Yearbook. All are divided by indexes of population derived for the major importing countries of each U.S. crop.

U.S. wheat and corn prices are from the I.M.F. International Financial Statistics tape. The soy price is from the Commodity Yearbook. All have been deflated by the U.S. CPI.

The R.O.W. wheat price is a price index for Australian wheat, denominated in U.S. dollars, and taken from the I.M.F. tapes. The corn price is a similar Argentine series from the same source. The soybean price is a series for Brazil, also from the I.M.F. tape. The soybean price series contains a small number of gaps. We have used simple interpolation to provide a complete series.

Additional exogenous variables used in the 2SLS estimations include real US GNP, plus various price series for production inputs taken from U.S.D.A. Agricultural Prices.

Each variable is converted to log form for the regressions.

NOTES

1. This somewhat cumbersome procedure was necessitated because no single SAS procedure exactly fit our needs.

REFERENCES

Abbott, Philip C. and Philip L. Paarlberg. Impacts of the 1980 Suspension of U.S.Grain Sales to the U.S.S.R. Station Bulletin No. 504. Department of Agricultural Economics, Purdue University, West Lafayette, Indiana, October 1986.

Adelman, Irma and Sherman Robinson. "U.S. Agriculture in a General Equilibrium Framework: Analysis with a Social Accounting Matrix." American Journal of Agricultural Economics 68(December 1986):1196-1207.

Anderson, Kym and Rodney Tyers. "European Community Grain and Meat Policies: Effects on International Prices, Trade, and Welfare." European Review of Agricultural Economics 11(1984):367-94.

Armington, Paul. "A Theory of Demand for Products Distinguished by Place of Production." IMF Staff Papers 16(July 1969):179-99.

Batten, Dallas and Michael T. Belongia. "The Recent Decline in Agricultural Exports: Is the Exchange Rate the Culprit?" Economic Review, Federal Reserve Bank of St. Louis 66(October 1984):5-14.

Becketti, Sean and Craig Hakkio. "How Real is the Real Exchange Rate?" Paper presented to the International Economics Workshop, Department of Economics and Business, North Carolina State University, Raleigh, NC, May 1987.

Belongia, Michael T. "Estimating Exchange Rate Effects on Exports: A Cautionary Note." Economic Review, Federal Reserve Bank of St. Louis 68(January 1986):5-16.

Blandford, David. "Modeling the Linkage Between Domestic and International Markets." Cornell Agricultural Economics Staff Paper No. 86-24. Department of Agricultural Economics, Cornell University, Ithaca, New York, August 1986.

Boughton, James, Richard Haas, Paul R. Masson, and Charles Adams. "Effects of Exchange Rate Changes in Industrial Countries." in International Monetary Fund Staff Studies for the World Economic Outlook. Washington (July 1986):115-49.

Branson, William. "The Trade Effects of the 1971 Currency Realignments." Brookings Papers on Economic Activity (1972)15-69.

Bredahl, Maury E., William Meyers, and Keith J. Collins. "The Elasticity of Foreign Demand for U.S. Agricultural Products: The Importance of the Price Transmission Elasticity." American Journal of Agricultural Economics 61(February 1979):58-63.

Caves, Richard E. and Thomas A. Rugel. "New Evidence on Competition on the Grain Trade." Food Research Institute Studies 18(1983 No. 3):261-74.

Chambers, Robert and Richard E. Just. "Effects of Exchange Rate Changes on U.S. Agriculture: A Dynamic Analysis." American Journal of Agricultural Economics (February 1981):32-46.

Clark, Peter B. "The Effects of Recent Exchange Rate Changes on the U.S. Trade Balance," in Clark et al., eds., The Effects of Exchange Rate Adjustments. Washington: U.S. Department of Treasury, 1974.

Collins, H. Christine. "Price and Exchange Rate Transmission." Agricultural Economics Research 32(October 1980):50-55.

Cox, Michael. "A New Alternative Trade-Weighted Dollar Exchange Rate Index." Economic Review, Federal Reserve Bank of Dallas (September 1986):20-28.

Deardorff, Alan, and Robert Stern. The Michigan Model of World Production and Trade. Cambridge: MIT Press, 1986.

Dornbusch, Rudiger. "Exchange Rates and Fiscal Policy in a Popular Model of International Trade." American Economic Review 65(December 1975):859-71.

________________. "Exchange Rates and Prices." National Bureau of Economic Research Working Paper No. 1769, December 1985.

Dutton, John. "Effects of Exchange Rates on Domestic Agricultural Prices." Department of Economics and Business, North Carolina State University, June 1987.

Dutton, John, and Thomas Grennes. Measurement of Effective Exchange Rates Appropriate for Agricultural Trade. Economics Research Report No. 51, Department of Economics and Business, North Carolina State University, Raleigh, N.C., November 1985.

_________________________. "Alternative Measures of Effective Exchange Rates for Agricultural Trade." European Review of Agricultural Economics, forthcoming.

Edwards, Clark. "The Exchange Rate and U.S. Agriucltural Exports." Agricultural Economics Research 39(Winter 1987):1-12.

Edwards, Sebastian. "Economic Liberalization and the Equilibrium Exchange Rate in Developing Countries." NBER Working Paper No. 2179, March 1987.

Evans, Paul. "Is the Dollar High Because of Large Budget Deficits?" Journal of Monetary Economics 18(November 1986):227-51.

Feldstein, Martin. "The Budget Deficit and the Dollar." NBER Working Paper No. 1898, April 1986.

Feldstein, Martin, and Phillippe Barchetta. "How Far Has the Dollar Fallen?" NBER Working Paper No. 2122, March 1987.

Frankel, J. A. "Expectations and Commodity Price Dynamics: The Overshooting Model." American Journal of Agricultrual Economics 68(1986):344-48.

Frenkel, Jacob and Michael Mussa. "Asset Markets, Exchange Rates, and the Balance of Payments: The Reformulation of Doctrine," in R. Jones and P. Kenen, eds., Handbook of International Economics, 2(1985), Amsterdam: North Holland.

Gardiner, Walter and Praveen Dixit. Price Elasticity of Export Demand: Concepts and Estimates. USDA, ERS, Foreign Agricultural Economic Report No. 228, February 1987.

Gardner, Bruce. "On the Power of Macroceonomic Linkages to Explain Events in U.S. Agriculture." American Journal of Agricultural Economics 63(December 1981):871-78.

Ginsburgh, Victor A. "A Further Note on the Derivation of Quarterly Figures Consistent with Annual Data." Applied Statistics 22(1973):368-74.

Goldstein, Morris and Mohsin Khan. "Income and Price Effects in Foreign Trade," in R.W. Jones and P.B. Kenen, eds. Handbook of International Economics, Vol. II(1985), New York: Elsevier Science.

Grennes, Thomas. "Domestic Commodity Programs, the Value of the Dollar, and United States Agricultural Exports." American Enterprise Institute Working Paper, Washington, March 1987.

Grennes, Thomas and John S. Lapp. "Neutrality of Inflation in the Agricultural Sector." Journal of International Money and Finance 5(June 1986):231-43.

Grennes, Thomas, Paul R. Johnson, and Marie C. Thursby. Economics of World Grain Trade, New York: Praeger, 1978.

Hakkio, Craig. "Does the Exchange Rate Follow a Random Walk? A Monte Carlo Study of Four Tests for a Random Walk." Journal of International Money and Finance 5(June 1986):221-29.

Henneberry, David, Mark Drabenstott, and Shida Henneberry. "A Weaker Dollar and U.S. Farm Exports: Coming Rebound or Empty Promise." Federal Reserve Bank of Kansas City, Economic Review 72(May 1987):22-36.

Henneberry, David, Shida Henneberry, and Luther Tweeten. "The Strength of the Dollar: An Analysis of Trade-Weighted Foreign Exchange Rates with Implications for Agricultural Trade." Agribusiness 3(Summer 1987):189-206.

Houthakker, Hendrik, and Stephen Magee. "Income and Price Elasticities in World Trade." Review of Economics and Statistics 51(1969):111-25.

Hsieh, David. "The Determination of the Real Exchange Rate." Journal of International Economics 12(1982):355-62.

Johnson, Paul R., Thomas Grennes, and Marie C. Thursby. "Devaluation, Foreign Trade Controls, and Domstic Wheat Prices." American Journal of Agricultural Economcis 59(November 1977):619-27.

Kost, William. "Effects of an Exchange Rate Change on Agricultural Trade." Agricultural Economics Research 28(July 1976):99-106.

Krissoff, Barry and Art Morey. The Dollar Turnaround and U.S. Agricultural Exports. USDA-ERS Staff Report Number AGES861128, December 1986.

Longmire, Jim and Art Morey. Strong Dollar Dampens Demand for U.S. Farm Exports. ERS-USDA Foreign Agricultural Economic Report No. 193, Washington: December 1983.

Magee, Stephen. "Prices, Income, and Foreign Trade: A Survey of Recent Economic Studies," in P. B. Kenen, ed., International Trade and Finance: Frontiers for Research. Cambridge: Cambridge University Press, 1975.

Meilke, Karl. "A Comparison of the Simulation Results from Six International Trade Models." Working Paper WP87/3, Department of Agricultural Economics and Business, University of Guelph, February 1987.

Mitchell, Donald O. and Ronald Duncan. "Market Behavior of Grain Exporters." World Bank Research Observer 2(January 1987):3-22.

Mussa, Michael. "Nominal Exchange Rate Regimes and the Behavior of Real Exchange Rates: Evidence and Implications." Carnegie-Rochester Conference Series on Public Policy 25(Autumn 1986).

Obstfeld, M. "Overshooting Agricultural Commodity Markets and Public Policy: Discussion." American Journal of Agricultural Economcis 68(May 1986):420-21.

Orcutt, Guy. "Measurement of Price Elasticities in International Trade." Review of Economics and Statistics 32 (May 1950):117-32.

Ott, Mack. "The Dollar's Effective Exchange Rate: Assessing the Impact of Alternative Weighting Schemes." Review Federal Reserve Bank of St. Louis (February 1987):5-14.

Pauls, B. Dianne. "Measuring the Foreign Exchange Value of the Dollar." Federal Reserve Bulletin 73(June 1987):411-22.

Pigott, Charles and Vincent Reinhardt. "The Strong Dollar and U.S. Inflation." Federal Reserve Bank of New York, Quarterly Review, August 1985.

Richardson, J. David. "Some Empirical Evidence on the Law of One Price." Journal of International Economics 8(August 1978):341-51.

Roe, Terry, Mathew Shane, and De Huu Vo. Price Responsiveness of World Grain Markets. ERS U.S. Department of Agriculture, Technical Bulletin, Number 1720, Washington, D.C.,June 1986.

Roningen, Vernon. "Trade Liberalization: Results from World Trade Models." Paper presented to the annual meeting of the IATRC, CIMMYT, Mexico, December 1986.

Schmitz, Andrew, Alex McCalla, Donald O. Mitchell, and Colin A. Carter. Grain Export Cartels. Cambridge: Ballinger, 1981.

Scobie, Grant. Food Subsidies in Egypt: Their Impact on Foreign Exchange and Trade. International Food Policy Research Institute Research Report No. 40, Washington, August 1983.

Seely, Ralph. "Price Elasticities from the IIASA World Agricultural Model." ERS-USDA, ERS Staff Report No. AGES850418, Washington, May 1985.

Sharples, Jerry A. "The Elasticity of Demand for Agricultural Exports: ERS Estimates." Mimeo, March 1985.

Sharples, Jerry A. and Philip L. Paarlberg. "Japanese and European Community Agricultural Trade Policies: Some U.S. Strategies." IED Staff Report, ERS-USDA, Washington, July 1982.

Spitaeller, E. "Short-Run Effects of Exchange Rate Changes on Terms of Trade and the Trade Balance." IMF Staff Papers 27(1980):320-48.

Stamoulis, Kostas and Gordon Rausser. "Overshooting of Agricultural Prices." Department of Agricultural and Resource Economics, University of California, Berkeley, mimeo, January 1987.

Stern, Robert M., Jonathan Francis, and Bruce Schumacher. Price Elasticities in International Trade: An Annotated Bibliography. London: Macmillan, 1976.

Stevens, Guy V. G., et al. The U.S. Economy in an Interdependent World: A Multicountry Model. Washington: Board of Governors of the Federal Reserve System, 1984.

Stockman, Alan. "The Equilibrium Approach to Exchange Rates." Federal Reserve Bank of Richmond, Economic Review 73(March/April 1987):12-30.

Thompson, Robert. A Survey of Recent U.S. Developments in International Agricultural Trade Models. ERS-USDA. Bibliographics and Literature of Agriculture No. 21, Washington, September 1981.

Thurman, Walter. "Endogenous Testing in a Supply and Demand Framework." Review of Economics and Statistics 68(November 1986):638-46.

Thursby, Jerry G. and Marie C. Thursby. "How Reliable Are Simple, Single Equation Specifications of Import Demand?" Review of Economics and Statistics, February 1984.

Thursby, Marie. "Strategic Models, Market Structure, and State Trading: An Application to State Trading," in Robert Baldwin, ed., Trade Policy Issues and Empirical Analysis. Chicago: University of Chicago Press, forthcoming.

Tyers, Rod and Kym Anderson. "Imperfect Price Transmission and Implied Trade Elasticities in a Multicommodity World Food Model." Paper presented to the IATRC Symposium on Elasticities in International Agricultural Trade, Dearborn, Michigan, July 31 - August 1, 1987.

U.S. Department of Agriculture. Agricultural Outlook, Washington, monthly.

Wilson, John F. and Wendy E. Takacs. "Differential Response of Price and Exchange Rate Influences in the Foreign Trade of Selected Industrial Countries." Review of Economics and Statistics 61(May 1979):269-79.

Chapter 5

Irma Adelman and Sherman Robinson

Macroeconomic Shocks, Foreign Trade, and Structural Adjustment: A General Equilibrium Analysis of the U.S. Economy, 1982-1986

INTRODUCTION

Since 1980, there have been major shifts in macroeconomic balances in the U.S. economy. The government deficit has greatly increased, accompanied by increased foreign borrowing and large balance of trade deficits. There have also been major shifts in relative prices, the exchange rate, and the sectoral structure of production, exports, imports, and domestic demand. The macro policy mix in the Reagan years has led to an effective revaluation of the dollar and to high U.S. real interest rates. The revaluation led to a significant decrease in the average relative price of tradables to non-tradables (the real exchange rate), which shifted incentives away from exporting, toward importing and the production of non-tradables.

In effect, the changes in relative prices induced by the macro swings constituted an industrial policy, shifting resources away from sectors producing tradables (exports and import substitutes) toward non-tradables. The agricultural sectors were also affected, both by the shifts in relative prices and by the high real interest rates. The revaluation hurt agricultural exports and the high real interest rates were capitalized in declining land prices, leading to severe financial problems for farmers in debt. Government policies were instituted to support agricultural prices, coupled with policies designed to restrict agricultural supply.

In this paper, we use a computable general equilibrium (CGE) model of the U.S. economy to analyze the impact of the swings in macro balances on the structure of relative prices, production, trade, income, and demand. The model is designed to focus on foreign trade issues, incorporating sectoral demand elasticities for imports and supply elasticities for exports. One issue we consider is the impact of different assumptions about these elasticities on the structural adjustments induced by the changes in macro balances. In the next section, we present a summary of the CGE model. We next discuss calibration of the model for 1982 and a base solution for 1986. We then analyze experiments in which we consider the impact of alternative macro policies designed to finance the increase in government expenditure observed during the period without recourse to increased foreign borrowing.

A CGE MODEL OF THE U.S. ECONOMY

Our CGE model is in the tradition of models developed for the analysis of issues of trade policy.[1] The model equations describe the supply and demand behavior of the various economic actors across markets for factors and commodities, including exports and imports. The model is neoclassical and Walrasian in spirit, solving for a set of relative prices (including the exchange rate) that achieve flow equilibria in the various markets. While there are many examples of CGE models that capture "structuralist" rigidities in the economy, the model we use here stays close to the neoclassical paradigm. For example, the model incorporates labor supply functions and, in all the experiments reported in this paper, achieves a full-employment equilibrium, with wages adjusting to clear the labor market.

The model contains ten sectors, including three agricultural sectors, five industrial sectors (including construction), and two service sectors. Sectoral production functions are all Cobb-Douglas in labor and capital. In two sectors, grains and other agriculture, cultivated land is also included as a factor. The total supply of cultivated land is fixed, but is assumed to be able to be allocated freely to either crop. The model thus solves for a single equilibrium land rental rate. Sectoral capital stocks are assumed to be immobile, so that model solutions generate differential rental rates across sectors. Labor is assumed to be completely mobile across sectors, and the model solves for a single equilibrium wage. The demand for intermediate inputs is given by fixed input-output coefficients.[2]

On the demand side, the model includes the following actors who receive income and demand goods: households, government, capital account, and the rest of the world. There are three categories of households classified by income level who receive income from wages, profits, rents, and transfers. They in turn save (according to fixed average savings rates) and then allocate their consumption expenditure across goods according to a simple linear expenditure system. Aggregate government expenditure on goods is specified exogenously (in real terms) and is allocated across sectors according to fixed shares. The government receives income from taxes (direct and indirect) which it then spends on goods and transfers to households. Government savings (the deficit) is determined residually as receipts minus expenditures. The capital account deals only in flows from current income, collecting savings from all sources (private, government, and foreign) and spending it on investment goods. The model is static in that the sectoral and aggregate capital stocks are fixed, and the investment flow is not "installed" or incorporated into sectoral capital stocks as part of an experiment.[3]

On the import side, the model specifies product differentiation between imports and domestically produced goods in the same sector. Demanders purchase a "composite" commodity in each sector which is a constant elasticity of substitution (CES) aggregation of domestically produced and imported goods.[4] The effect is that import demand is a function of the ratio of the price in domestic currency of the import (PM) to that of the domestic good in the same sector (PD). On the export side, suppliers are assumed to have different production functions for goods sold on the domestic and export markets. Using factor inputs, they produce a "composite" commodity which can then be transformed into goods intended for exporting versus those for the domestic market according to a constant elasticity of transformation (CET) function. Given the assumption of profit maximization, the ratio of export goods to goods for the domestic market in each sector is a function of the relative price in domestic currency of exports (PE) and domestic sales (PD). In effect, each sector is a two-product firm with a separable production function. The determination of the level of aggregate production is based on the producer price of the composite commodity (PX), while the composition of supply to the export and domestic markets depends only on the relative prices in the two markets (PE/PD).

With respect to the world market, we retain the standard "small country" assumption. Sectoral world prices of exports and imports (PWE and PWM) are assumed to be fixed exogenously and are independent of the volume of exports and imports. While such a specification would not be adequate if we were focusing on, say, export incentive policies in grains, it is acceptable given that we are concerned in the experiments with the structural impact of swings in macroeconomic balances.

The effect of this trade specification is partially to insulate the domestic price system from world prices. In a model in which all goods are tradable and are perfect substitutes with foreign goods, domestic relative prices are completely determined by world prices. By contrast, in this model, all domestically produced goods sold on the domestic market are only imperfectly substitutable with goods either bought from or sold to the rest of the world. This specification has proven to yield much more realistic behavior than a model incorporating perfect substitutability and is widely used in CGE models focusing on international trade.

The model incorporates a number of different prices in each sector. On the demand side, the price of the composite good (P) corresponds to a retail sales price, and is a weighted average of the domestic currency prices of imports (PM) and domestic goods sold on the domestic market (PD). On the supply side, the producer price (PX) represents an average of the prices in domestic currency of goods sold on the domestic market (PD) and exports (PE). The domestic prices of imports and exports are related to world prices by the following equations:

$$PE = (1 + TE) \quad EXR \cdot PWE$$
$$PM = (1 + TM) \quad EXR \cdot PWM$$

where TE and TM are export subsidies and import tariffs, EXR is the exchange rate, and PWE and PWM are the world prices of exports and imports.[5]

Since the model only determines relative prices, some price must be chosen as numeraire. We chose an aggregate index of domestic prices (PD) as numeraire, using base-year output weights. In effect, we are fixing the average price of nontradables in the model. Thus, when we solve for the equilibrium exchange rate (EXR), we are effectively solving for the relative prices of tradables to nontradables in the domestic economy, or the real exchange rate. We chose this index as numeraire to facilitate interpretation of the equilibrium exchange rate solved in the model. While the numeraire does not correspond exactly to any standard price index (e.g., producer price index or consumer price index), the differences among the various solution aggregate price indices in our experiments was very small. The important point to note is that the model does not incorporate inflation, so all results are effectively measured against 1982 base prices.

The model focuses on flow equilibria and does not include any asset markets or money. It does, however, incorporate the major macroeconomic aggregate balances:

$$
\begin{aligned}
Z &= SH + SG + F \\
SG &= T - G \\
F &= M - E
\end{aligned}
$$

where Z is aggregate investment, SH is total private savings, SG is government savings, F is foreign savings (the balance on current account), T is total government revenue, G is government expenditure, M is aggregate imports, and E is aggregate exports. How a CGE model achieves balance among these macro aggregates in equilibrium defines the model's "macro closure." Issues of macro closure have been much discussed in the literature on CGE models.[6] For our analysis, however, the issue is straightforward.

Given our assumption of full employment, there can be no significant feedback from macro disequilibrium to employment and aggregate output. The model is Walrasian, not Keynesian. We are focusing on the impact of changes in the composition of these macro aggregates, and so essentially set them exogenously.

We assume that aggregate government expenditure on goods is fixed exogenously in real terms. Government revenue is determined by a variety of taxes, given fixed average tax rates. Government savings (the deficit) is thus determined endogenously, as a residual. Foreign savings (the balance on current account or the balance of trade in goods and services, including factor services) is set exogenously in world market prices. Its value in domestic currency depends on the equilibrium exchange rate. Private savings are generated by using fixed average savings rates for corporate and household income. Aggregate investment is determined by summing all savings. There is no independent investment function and no interest rate variable, so investment is essentially "savings driven."

In the experiments reported below, we vary foreign savings by changing the exogenously specified balance on current account. In this case, the real exchange rate (EXR) must adjust to generate the new equilibrium levels of imports and exports. We also vary the government deficit by changing the exogenously specified average tax rates on corporate and household income. In both cases, there are major changes in aggregate savings and hence investment. Since the model does not include interest rates or asset markets, we are effectively specifying experiments with complete "crowding out" or "crowding in" of investment. This treatment is adequate given our focus on examining the structural implications of swings in macro aggregates. We are not seeking to explain the process of macro adjustment.[7]

Our model has a close resemblance to another CGE model of the U.S. developed by Hertel and Tsingas (1987) --henceforth referred to as the H-T model-- which is also being used to analyze issues related to U.S. agricultural policy. The H-T model is larger than ours and has a much more detailed specification of the agricultural sector.[8] Their specification of agricultural technology is also more elaborate, incorporating partial cross-elasticities of substitution between land, capital, labor, and fertilizer. They also include more tax and subsidy policy instruments relating to agriculture. Their focus is on public finance issues such as efficiency losses arising from various tax and subsidy schemes, so the additional technological and institutional detail is needed.

The specification and parameter values of the H-T model indicate that it has a shorter run focus than our model. For example, they specify relatively low elasticities of substitution among factor inputs in the agricultural sectors and also assume that agricultural labor cannot migrate to other sectors. On the trade side, they specify relatively low trade substitution elasticities between imports and domestic goods and also specify very low export demand elasticities (less than one in some cases). By contrast, our Cobb-Douglas production technology assumes substitution elasticities among factor inputs of one and we assume agricultural labor is freely mobile across sectors. In general, we specify higher trade substitution elasticities and also adopt the "small country" assumption, with infinite elasticities of demand for U.S. exports and of foreign supply of imports to the U.S.

On the macro side, H-T specify that aggregate investment is fixed in real terms, while in our model it is determined by aggregate savings. Again, they appear to have a shorter run focus for which such a specification is reasonable, and also it is convenient to fix real investment when making welfare comparisons with comparative static experiments, which is their major focus. Their model is based on data for 1977, before the major exchange rate and interest rate movements, while our base year is 1982 and our analysis covers the period 1982-1986.

1982 DATA AND 1986 BASE SOLUTION

Table 5.1 provides a social accounting matrix, or SAM, for the U.S. economy in 1982.[9] The SAM in Table 5.1 shows the macro aggregates and the flows among the various actors in the model, grouping the income and expenditure accounts of each actor into a square matrix. The sectoral and household accounts are aggregated for presentation; the full SAM includes the input-output accounts and different types of households. A SAM provides the underlying data framework for a CGE model in much the same way that the national income and product accounts underlie macro models.[10]

The model disaggregates households into three groups ranked by income: the bottom 40 percent of households, the middle 40 percent, and the top 20 percent. The model includes ten sectors (or "activities" in the SAM), and Table 5.2 provides sectoral detail on the structure of production, value added, and trade. The choice of sector aggregation partly reflects an attempt to group sectors with similar trade characteristics. For example, the agricultural sectors were grouped into one with a low trade share (dairy and meat), one with high exports (grains), and one with significant imports (other agriculture). Table 5.2 also gives the assumed elasticities of import substitution and export transformation by sector. These data indicate the importance of trade in different sectors and their responsiveness to changes in relative prices.

The parameters of the model are calibrated so that the base-year data for 1982 represent an equilibrium solution. Most of the parameters are computed using base-year shares from the SAM.

The various trade elasticities are "guesstimates" based on a literature survey of scattered econometric work. No original econometric estimation has been done for this model. Units are chosen so that all product prices (including imports, exports, and domestic sales) and the exchange rate equal one in the base year.

Given the 1982 solution, we then generate a solution for 1986 by specifying 1986 values for the exogenous variables in the model, including: total labor force, sectoral capital stocks, total acreage planted, real government expenditure (including its sectoral composition), average tax rates, the current account balance, and world prices of exports and imports. Most of the parameters in the model were assumed unchanged, including those for the production functions, import aggregation functions, export transformation functions, and household expenditure functions. In the production functions, both the input-output coefficients and the level of total factor productivity were assumed constant (i.e., no total factor productivity growth). We also held the numeraire price index for domestic goods constant, so all exogenous variables were projected in terms of constant 1982 prices. World prices of exports and imports for 1986 were projected to move together, except for basic intermediates (which includes oil) in which there was a significant relative decline in the import price (an improvement in the international terms of trade).

The model solution for 1986 is given in Tables 5.3 to 5.7. In general, the solution agrees well with the preliminary data that are available for 1986. The various macro aggregates are within roughly a percent or two of the data, and the relative price movements in the solution agree with the scattered evidence available, including the projected 15.3 percent revaluation in the real exchange rate. Given the uncertainty in the preliminary 1986 data, there is little reason to work to refine the 1986 solution further until more detailed data become available.[11] In any case, sensitivity analysis with the model indicates that the basic results from the macro experiments reported below are very robust to variations in the 1986 base solution.

MACRO SHOCKS AND FULL-EMPLOYMENT INCOME

Between 1982 and 1986, the economy recovered from a recession. Real annual growth rates were 3.8 percent for GNP, 2.4

percent for civilian employment, and 5.1 percent for aggregate absorption (GNP + imports - exports).[12] During this period, however, imports rose dramatically and exports stagnated. Net exports of goods and services (the balance on current account in the national income and product accounts) moved from a surplus of 26.3 billion in 1982 to a deficit of 105.7 billion in 1986 (in current prices). During the period, the U.S. also benefited from a significant improvement in its international terms of trade, largely through the collapse of the world price of oil. Real imports thus grew faster than nominal imports, and the real net export balance in 1986 was -149.7 billion in 1982 dollars. In the same period, the total government deficit (including federal, state, and local government) rose from $111 billion to $143 billion. In 1980, at the beginning of the Reagan administration, the total government deficit was only $34.5 billion.

In this section, we analyze the economic impact of the macroeconomic shocks observed in the 1980s on the structural characteristics of production, employment, international trade, and the size distribution of income among households.[13] In analyzing the impact of these macro shocks, we consider a set of counterfactual questions which we simulate with the CGE model. The first question we ask is: "Ceteris paribus, had we achieved the same balance of trade surplus as in 1980 (the last Carter year), what would the effects have been?" This "foreign savings" experiment assumes no reliance on foreign borrowing to finance the deficit and is modelled by changing the exogenous balance of trade in goods and services to achieve the same trade balance in 1986 as obtained in 1980. The second question is: "Ceteris paribus, had the increase in government expenditures been financed by increasing taxes without generating a budget deficit, what would the effects have been?" This "government savings" experiment is modelled by increasing the exogenous tax rates on corporate income and the income of the richest households to achieve approximately the same total government deficit in 1986 as obtained in 1980. In a third "combination" experiment, we combine the first two and ask: "Ceteris paribus, had both foreign borrowing and the budget deficit been at 1980 levels, what would the full-employment U.S. economy have looked like in 1986?"

The results of these experiments are given in Tables 5.3 to 5.7. The foreign borrowing increased absorption by 4.3 percent and went mostly into investment, which was 22.5 percent higher than it would have been without the borrowing, thus avoiding a crowding-out effect due to the government deficit. Foreign borrowing also reduced the budget deficit by about $8 billion, since employment was 0.5 percent higher than without the borrowing, and hence social security taxes and household income taxes were about $4 billion more.[14] But the major effect of restricted foreign capital inflows is on the exchange rate: the 15 percent revaluation in exchange rate relative to 1982 which occurred in 1986 would not have taken place had we not borrowed from abroad. Exporters would not have suffered a loss in competitiveness and imports would have been $83 billion less than they were in 1986.

The revaluation of the real exchange rate had a major effect on the agricultural terms of trade, partly as a result of its dramatic effect on agricultural exports.[15] The experiment indicates that agricultural exports would have been about 20 percent higher if the real exchange rate had not risen. The exchange rate appreciation induced by the foreign finance of the budget deficit also increased agricultural imports dramatically: by 33 percent in dairy production, 15 percent in grains, and 30 percent in other agriculture.[16] Clearly, farmers were hurt by the increased foreign borrowing; the experiment indicates that their incomes were about $12 billion lower than they would have been had foreign savings not been used to finance the investment recovery. The actual payment to farmers in 1982 dollars under the government commodity programs was $13.8 billion.[17] In the aggregate, farmers were thus compensated by just about the right amount for the income losses they incurred as a result of the declines in their terms of trade. This, of course, does not mean that the distribution of the compensation matched the distribution of losses. The empirical evidence suggests it did not.[18]

Under the "foreign saving" experiment, the "rent" on land increases by 18 percent. That is, assuming full capitalization, the price of land would have been some 18 percent higher in 1986 if foreign capital inflows had been at their 1980 level. This computation assumes no change in the real interest rate. In fact, the real interest rose dramatically during this period, a rise that was required to attract the foreign savings. In 1986, the real interest rate was about 6.4 percent, around triple the "normal" long-term real rate.[19] At any degree of capitalization, these swings in the real interest rate inflicted much more serious declines in land values than occurred through the loss in land rental rates. In any case, the two effects worked in the same direction, dramatically lowering land prices and leading to financial problems for farmers who had borrowed against land.

Our second counterfactual experiment, the "government savings" experiment, in which the government deficit is set to zero and government expenditures are financed by corporate tax increases and by raising the tax rate on the richest households, essentially restores the exchange rate and the agricultural terms of trade and agricultural incomes to their 1986 values. The experiment has only minor effects on absorption, employment, wages, imports and exports as compared to their actual 1986 values. The major effect of the budget deficit on full-employment income appears to be that consumption is lower, in both absolute and relative terms, and investment is considerably higher than with the 1986 deficit. This experiment makes it clear that the major effect of the budget deficit per se is to crowd out investment. With a more conservative budget policy, investment would have been 17.5 percent higher and consumption 4 percent lower than in 1986.

The final macro experiment combines both counterfactuals, setting both the trade deficit and the budget deficit to their pre-Reagan values. The results are close to those of the "foreign savings" experiment in aggregate terms and in their incidence on farmers and farm production, and to the "government savings" experiment in composition of GNP between consumption and investment and in investment-related production.

The experiments suggest that the effects of the Reagan macro policies have been dominated by the import of capital from abroad. The effect on the U.S. economy has been a major revaluation of the real exchange rate, which imposed a heavy burden on exporters and farmers. Its effects on the rest of the world have been more mixed. On the one hand, we have syphoned savings out of the rest of the world, reducing their investment rate. On the other hand, by vastly increasing our imports over what they would have been with more conservative macro policies, we have generated export-led growth in our supplier countries.

SENSITIVITY TO TRADE ELASTICITIES

The ability of the economy to adjust to trade-related macro shocks depends on its ability to substitute domestic for foreign goods in demand and to shift between domestic and world markets in production. Our model assumes moderately elastic import substitution and export transformation elasticities (see Table 5.2 above), which are reasonable for the five-year time horizon of our experiments. One would expect these elasticities to be less in the short run. In any case, it is important to explore the sensitivity of the results to variations in these parameters.

Tables 5.8 and 5.9 report the results from an experiment in which we replicated the exogenous changes underlying the 1986 base run, but cut all the sectoral import substitution and export transformation elasticities in half. On average, the elasticities on both the export and import sides become less than one, indicating a major change in the economy's ability to adjust to a shock through changes in trade. Intuitively, restricting the ability of the economy to make a quantity adjustment should increase the observed price adjustment; and this is the case. The real revaluation in the half-elasticity experiment is 29 percent compared to 15 percent in the 1986 base run.

Intuition is less clear about what should happen to the aggregate volume of imports and exports, and to their sectoral composition. There are a number of different forces at work. At the sectoral level, the elasticity of demand for the composite good, the dependence on imported intermediate inputs, and the trade shares in both exports and imports are at least as important as the trade elasticities.[20] In addition, overall trade is affected by changes in the sectoral composition of aggregate production and demand. At the aggregate level, the result from this experiment is to increase the volume of foreign trade. The sum of exports and imports goes up by $44 billion (or 5 percent) relative to the 1986 base run total. But, even though trade elasticities are cut, the economy retains enough substitution possibilities in production and demand so that total real GNP is identical in the two runs.

In the case of agriculture, farmers would gain in a situation in which all trade elasticities are lower. After the change, their import substitution elasticities and the export transformation elasticity in grains remain above one. However, agricultural supply elasticities are lower than those in most other sectors since the agricultural sectors have two specific factors (land and capital) while other sectors have only capital as a specific factor. In all three agricultural sectors, in the new equilibrium, imports are lower and exports are higher than in the 1986 base run. With the higher relative demand and low supply elasticities, the agricultural terms of trade improve substantially (106.4 in the low-elasticity experiment compared to 101.9 in the 1986 base).

We did a number of experiments replicating the macro shock experiments, but with lower trade elasticities. In general, the results are consistent with the experiment reported above: the real exchange rate varied more, but the macro aggregates were unchanged. Also, the basic results with regard to agriculture remained. Agriculture gains from alternative macro policies in the 1982-1986 period that would have lowered the trade deficit, even with much lower trade substitution elasticities.

CONCLUSION

The experiments we have performed with the CGE model indicate the importance of general equilibrium linkages and of price effects in transmitting a macro shock through the economy. Attempts to isolate the impact of such shocks on a single sector using partial equilibrium analysis are likely to be misleading, as indicated by the results from our elasticity-sensitivity experiments. Trade policy, especially when it affects the exchange rate, reverberates throughout the economy, and is an area where general equilibrium analysis is especially necessary.

Our experiments also indicate that the impacts of changes in macroeconomic and trade policy at the sectoral level are sensitive to trade elasticities (in their own and other sectors). In our model, these elasticities are essentially "guesstimates," and are not based on econometric work. Our sensitivity analyses indicate the need for careful econometric estimates of these parameters.

The results from our particular macro experiments appear to be quite robust and yield a number of interesting policy conclusions. Agriculture suffered substantially from the macro policies actually followed in the 1981-1986 period. Through increased foreign borrowing, the economy as a whole had a higher growth rate and a higher investment rate, while maintaining consumption growth, than would have been possible if the increasing government deficit had been financed domestically. It also underwent a substantially larger adjustment in the structure of production and trade than would have otherwise occurred. The mechanism through which this adjustment was induced was a major revaluation in the real exchange rate, leading to major changes in domestic relative prices.

It is arguable whether the long run effects of the overvalued real exchange rate have been good or bad. In essence, what the high real exchange rate has accomplished is the equivalent of policies of structural adjustment through import liberalization, policies which have been advocated by many economists for developing countries. To the extent that the United States is incapable of implementing good industrial policies, the very high real exchange rate provided an alternative which has had an across-the-board effect of forcing U.S. producers in manufacturing to cut costs and become more competitive, and to reduce output in sectors in which it has no comparative advantage (although the damage to exporters was clearly excessive). Like many developing countries, the United States has borrowed in order to achieve this structural adjustment in a relatively painless way. As a result of the restructuring which has taken place, U.S. manufacturing and agriculture are both more competitive. They are now poised to take advantage of the decline in the real exchange rate that is now occurring.

The dangers of the policy of financing the restructuring of the U.S. economy by foreign borrowing are a debt overhang, the potential for greater macroeconomic instability, and less scope for independent adjustment of U.S. macro policies in the future. We have recently had a "soft landing" for the exchange rate devaluation accompanied by only moderate inflationary pressures so far. By borrowing, we have bought ourselves some time and an easier adjustment, but it is not clear how much longer we can continue to increase our foreign debt.

It is important to understand that the current trade deficit was brought about by U.S. macro policy choices and not by the policies of our trading partners. Indeed, it is difficult to see what they could have done about it, since they could not singly do anything to change our macro policies. However, the recent exchange rate devaluations signal a change. It will be increasingly difficult for the U.S. to finance its deficit through foreign borrowing, and there will be increasing pressure on the U.S. to change macro policy. The change also signals a major opportunity for U.S. producers of tradable goods, both exports and import substitutes. It would be a great pity, both for the U.S. and the world economy, if this opportunity were to be missed by a round of protectionist legislation, perhaps setting loose a full-scale trade war.

Table 5.1. Social Accounting Matrix for the United States, 1982

Receipts	Expeditures (billion dollars) Commodities 1	Activities 2	Factors (value added) 3	Indirect taxes 4	Employee compensation 5	Proprietors' income 6	Other property income 7	Enterprises 8	Households 9	Capital accounts 10	Government 11	Rest of the world 12	TOTAL
1. Commodities		2,892.4							1,984.9	414.9	650.5	348.4	6,291.1
2. Activities	5,961.7												5,961.7
3. Factors (value added)		2,810.5											2,810.5
4. Indirect taxes		258.8											258.8
Sum (GNP)		3,069.3											
5. Employee compensation			1,864.2										1,864.2
6. Proprietors' income			111.5										111.5
7. Other property income			470.7										470.7
Sum (national income)			2,446.4										
8. Enterprises			14.1				470.7				44.4		529.2
9. Households					1,612.9	111.5		439.3			362.0	-1.2	2,524.5
10. Capital account			358.8					29.2	135.5		-115.2	6.6	414.9
11. Government			-8.8	258.8	251.3			60.7	404.1			-24.4	941.7
12. Rest of the world	329.4												329.4
TOTAL	6,291.1	5,961.7	2,810.5	258.8	1,864.2	111.5	470.7	529.2	2,524.5	414.9	941.7	329.4	

Table 5.2. Sectoral Composition, Trade Shares, and Elasticities

Sector	Sectoral compostion (percent): Value added	Gross output	Exports	Imports	Trade shares (percent): Exports/XD	Imports/X	Elasticities: Import substitution	Export transformation
	1	2	3	4	5	6	7	8
1. Dairy	0.3	1.3	0.1	0.2	0.3	0.8	4.0	1.5
2. Grains	1.3	1.2	5.0	0.0	24.3	0.2	4.0	4.0
3. Other agriculture	1.0	0.8	0.5	1.4	3.7	9.5	4.0	1.5
Sum/average	2.6	3.3	5.6	1.6	9.4	3.5	4.0	1.5
4. Light consumer	6.6	10.9	7.9	12.9	4.3	6.6	2.0	3.0
5. Basic intermediate	9.7	14.8	16.1	34.3	6.4	12.9	3.0	3.0
6. Capital goods	6.1	8.5	23.0	18.0	15.9	11.8	1.2	3.0
7. Construction	5.4	6.7	0.0	0.0	0.0	0.0	0.9	1.5
8. Electronics	1.9	2.0	4.4	9.0	13.1	25.3	1.1	3.0
Sum/average	29.7	42.9	51.4	74.2	7.9	11.3	2.5	3.0
9. Trade and finance	15.8	13.8	5.5	0	2.4	0.0	0.2	0.6
10. Other service	51.9	40.0	37.5	24.2	5.5	3.3	0.2	0.6
Sum/average	67.7	53.8	43.0	24.2	3.9	1.7	0.2	0.6
Overall Sum/Average	100.0	100.0	100.0	100.0	7.6	7.1	1.7	1.8

Sources:

Columns 1-6: Data for 1982, XD is gross production, and X is domestic supply (production plus imports minus exports).

Columns 7-8: Model parameters.

Table 5.3. Gross National Products Accounts: Base Run and Experiments

	Base runs		Experiments*		
	1982	1986	Foreign savings	Government savings	Combination
	1	2	3	4	5
Real GNP**	---billions of dollars---		---------ratio to1986 (percent)--------		
GNP	3,078.6	3,578.9	99.7	100.6	100.3
Consumption	1,994.0	2,276.2	99.6	96.1	95.5
Investment	415.1	619.8	77.5	117.5	96.1
Government	650.5	763.9	100.2	99.9	100.2
Exports	348.4	314.3	140.5	102.6	143.6
Imports	329.4	395.4	98.0	102.2	100.3
			-----------billions of dollars-----------		
Balance of trade	19.0	-81.1	54.4	-81.6	43.8
Government deficit	-115.0	-134.9	-143.2	0.2	0.3
Real trade (1982 prices)			---------ratio to 1986 (percent)---------		
Exports	348.4	370.0	120.8	101.9	123.0
Imports	329.4	519.1	84.1	101.6	85.9
Absorption and employment					
Absorption	3,059.6	3,666.3	95.7	100.5	96.3
Employment	96.6	106.4	99.5	100.2	99.8

*In the experiments, the average domestic price level is kept at the 1982 level. World prices of exports and imports were set reflecting 1986 international terms of trade. Exports and imports in the first part of the table are valued at the equilibrium exchange rate times the world price.
**GNP accounts are in real 1982 average price level, but 1986 relative prices.

Sources:
- Column 1: Social accounting matrix.
- Column 2: Base solution of CGE model.
- Column 3: Balance of trade set to 1980 level.
- Column 4: Government deficit set to 1980 level.
- Column 5: Combination of foreign savings and government savings experiments.

Table 5.4. Composition of Gross National Product and Savings: Base Run and Experiments

	Base runs		Experiments*		
	1982	1986	Foreign savings	Government savings	Combination
	1	2	3	4	5
	percent				
Gross national product shares					
Consumption	64.8	63.7	63.5	60.9	60.6
Investment	13.5	17.3	13.5	20.2	16.6
Government	21.1	21.3	21.5	21.2	21.3
Balance of trade	0.6	-2.3	1.5	-2.3	1.5
TOTAL	100.0	100.0	100.0	100.0	100.0
Investment shares					
Foreign savings	1.6	19.5	-6.1	16.6	-4.9
Private savings	126.1	102.3	135.9	83.4	104.9
Government savings	-27.7	-21.8	-29.8	0.0	0.0
TOTAL	100.0	100.0	100.0	100.0	100.0

*In the experiments the average domestic price level is kept at the 1982 level. World prices of exports and imports were set reflecting 1986 international terms of trade. Exports and imports in the first part of the table are valued at the equilibrium exchange rate times the world price.

Sources:
Column 1: Social accounting matrix.
Column 2: Base solution of CGE model.
Column 3: Balance of trade set to 1980 level.
Column 4: Government deficit set to 1980 level.
Column 5: Combination of foreign savings and government savings experiments.

Table 5.5. Price and Income Indices: Base Run and Experiments

Indices, 1982=100	Base runs 1982	1986	Experiments* Foreign savings	Government savings	Combination
	1	2	3	4	5
Agricultural terms of trade	100.0	101.9	108.9	101.4	108.2
Exchange rate	100.0	84.9	98.8	85.5	99.1
Domestic import price (PM)	100.0	79.8	92.8	80.3	93.1
Domestic export price (PE)	100.0	84.9	98.8	85.5	99.1
Producer price (PX)	100.0	99.1	99.9	99.1	100.0
Domestic market price (PD)	100.0	100.0	100.0	100.0	100.0
Composite good price (P)	100.0	98.5	99.4	98.5	99.4
Cost of living	100.0	99.8	100.2	99.7	100.0
Wage	100.0	106.7	106.4	106.9	106.6
Land rent	100.0	102.3	120.7	101.0	118.8
Household income					
Poorest 40 percent	100.0	111.2	111.0	108.0	107.8
Middle 40 percent	100.0	114.0	113.4	111.9	111.4
Richest 20 percent	100.0	113.7	113.3	110.2	109.8

*In the experiments the average domestic price level is kept at the 1982 level. World prices of exports and imports were set reflecting 1986 international terms of trade. Exports and imports in the first part of the table are valued at the equilibrium exchange rate times the world price.

Sources:
- Column 1: Social accounting matrix.
- Column 2: Base solution of CGE model.
- Column 3: Balance of trade set to 1980 level.
- Column 4: Government deficit set to 1980 level.
- Column 5: Combination of foreign savings and government savings experiments.

Table 5.6. Sectoral Gross Output: Base Data and Experiments

	Base runs		Experiments*		
	1982	1986	Foreign savings	Government savings	Combination
	1	2	3	4	5
	billions of dollars (1982)		ratio to 1986 (percent)		
Dairy	77.3	85.4	100.7	98.8	100.0
Grains	72.0	70.7	103.0	99.6	102.8
Other agriculture	46.8	47.2	105.0	99.6	103.8
Light consumer	647.4	725.9	100.5	99.0	99.5
Basic intermediate	877.9	929.5	104.0	101.8	106.0
Capital goods	504.0	597.3	99.4	107.0	107.3
Construction	399.0	531.8	86.8	110.0	97.5
Electronics	117.1	131.7	100.7	102.4	103.4
Trade and finance	820.0	938.0	98.7	99.5	98.1
Service	2,381.7	2,759.6	100.2	98.9	99.1

*In the experiments the average domestic price level is kept at the 1982 level. World prices of exports and imports were set reflecting 1986 international terms of trade. Exports and imports in the first part of the table are valued at the equilibrium exchange rate times the world price.

Sources:

Column 1: Social accounting matrix.
Column 2: Base solution of CGE model.
Column 3: Balance of trade set to 1980 level.
Column 4: Government deficit set to 1980 level.
Column 5: Combination of foreign savings and government savings experiments.

Table 5.7. Index of Producer Prices: Base Data and Experiments

Indices, 1982 = 100	Base runs 1982	Base runs 1986	Experiments* Foreign savings	Experiments* Government savings	Experiments* Combination
	1	2	3	4	5
Dairy	100.0	102.6	107.5	102.1	106.9
Grains	100.0	97.8	109.7	97.3	108.8
Other agriculture	100.0	103.0	108.5	102.6	107.9
Light consumer	100.0	99.4	101.2	99.2	101.0
Basic intermediate	100.0	92.8	95.0	93.6	95.7
Capital goods	100.0	100.4	100.9	100.8	101.4
Construction	100.0	100.1	100.1	100.8	100.9
Electronics	100.0	100.4	101.0	100.7	101.4
Trade and finance	100.0	102.1	101.7	102.1	101.7
Other service	100.0	99.6	99.8	99.3	99.4

*In the experiments the average domestic price level is kept at the 1982 level. World prices of exports and imports were set reflecting 1986 international terms of trade. Exports and imports in the first part of the table are valued at the equilibrium exchange rate times the world price.

Sources:
- Column 1: Social accounting matrix.
- Column 2: Base solution of CGE model.
- Column 3: Balance of trade set to 1980 level.
- Column 4: Government deficit set to 1980 level.
- Column 5: Combination of foreign savings and government savings experiments.

Table 5.8. Trade Elasticity Experiment, Sectoral Results (1986 Ratios to 1982 Base Run, Percent)

Sector	Output		Imports		Exports		Producer price		Retail price	
	Base	Half elasticity	Base	Half elasticity	Base	Half elasticity	Base	Half elasticity	Base	Half elasticity
Dairy and meat	110.5	111.1	235.4	203.1	83.1	88.6	102.6	105.5	102.4	105.2
Grains	98.2	99.6	216.7	217.8	55.9	57.0	97.8	103.3	100.8	109.1
Other agriculture	100.9	104.1	225.2	193.3	76.0	82.9	103.0	105.8	101.1	103.2
Light consumer	112.1	112.7	144.7	137.6	81.9	88.4	99.4	99.7	98.8	99.0
Basic intermediate	105.9	101.7	190.1	224.0	96.8	95.5	92.8	90.6	88.6	85.7
Capital goods	118.5	121.3	155.9	148.5	100.8	112.8	100.4	99.8	100.7	100.6
Construction	133.0	131.6			104.8	109.5	100.1	99.2	100.1	99.2
Electronics	112.5	115.5	143.6	138.2	80.5	90.3	100.4	99.9	98.0	96.7
Trade	114.4	114.3			102.1	105.2	102.1	101.9	102.5	102.4
Services	115.9	115.9	120.6	119.3	129.5	132.6	99.6	99.3	100.1	100.0
TOTAL	114.7	115.5	157.6	165.7	106.8	111.1	99.9	98.6	99.4	98.0

Table 5.9. Trade Elasticity Experiment, Aggregate Results (1986 Ratios to 1982 Base Run, Percent)

	Experiment:	
Variable	1986 Base run	Half elasticity
GNP	116.3	116.3
Consumption	114.2	113.9
Investment	149.3	146.1
Agricultural terms of trade	101.9	106.4
Exchange rate	84.9	78.1
Real wage	106.9	100.0
Land rent	102.3	112.5

NOTES

1. Our particular model is close in spirit to a model described in Condon, Robinson, and Urata (1985). Related models are discussed in detail in Dervis, de Melo, and Robinson (1982). Dixon and Parmenter (1986) have adapted their CGE model of Australia (called ORANI) to examine issues of structural adjustment similar to those we consider, although the models differ substantially. See also the survey by Shoven and Whalley (1984).

2. This specification of the production technology is quite simple, especially for the agricultural sectors. It represents a first step and it is clearly worthwhile to develop the specification further.

3. In fact, the model identifies investment by sector of destination and then converts it into demand for investment goods by sector of origin using fixed capital composition coefficients. Dynamically, this is a "putty-clay" formulation. In the comparative static experiments in this paper, it is more like "concrete."

4. This formulation follows Armington (1969). The implications of this specification within neoclassical trade theory and the effect on the behavior of CGE models is explored in de Melo and Robinson (1981, 1985, 1986).

5. In the 1982 base data, there are no export subsidies, so TE is set to zero in every sector. The tariffs are very low, averaging 2-3 percent.

6. For a survey of these issues in models of developing countries, see Robinson (1986) and Adelman and Robinson (1988).

7. It is interesting to compare our results, which assume a full-employment economy, with those from demand-driven multiplier models, such as Henry and Schluter (1985), which assume no supply constraints.

8. They use a linearization technique for solution, making a larger model feasible but introducing approximation errors.

9. The table is taken from Robinson and Roland-Holst (1987). There are some minor inconsistencies between the SAM and the base data used in the model arising from some simplifications in the model, including the treatment of tariffs and the consolidation of some transfers. The SAM is based on data for 1982, including an input-output table at the 528 sector level, provided by Engineering Economics Associates, Berkeley, California and reconciles exactly with the published national income and product accounts for 1982. See U.S. Department of Commerce (1984). Recent revisions to the macro data have not been taken into account.

10. A SAM can also provide a framework for multiplier analysis similar to input-output models. See Adelman and Robinson (1986) and Robinson and Roland-Holst (1987) who analyze various kinds of multiplier linkages with the same 1982 U.S. data set.

11. Since completing this paper, we have further refined the data base, making a number of changes. In particular, the very small change in the agricultural terms of trade shown in Table 5.5 has been revised, with later data indicating a fall over the 1982-86 period.

12. All data reported in this paragraph, and in any discussion of actual data below, come from U.S. Government (1987), the Economic Report of the President, 1987. There are some minor discrepancies between these data and 1982 data used in the model because of later revisions not taken into account in the input-output accounts used in the model.

13. See Freebairn, Rausser, and de Gorter (1983) for an analysis of the impact of U.S. monetary policy on agriculture.

14. The rest of the change in the deficit is due to the fact that official capital outflows (which appear as an exogenous government expenditure in the model) are fixed in foreign currency, and their value in domestic dollars increased with the devaluation.

15. There are indirect effects as well, since the structure of trade also changes in the nonagricultural sectors. See Schuh (1974) and Chambers and Just (1982) for a discussion of the direct effects of exchange rate changes on agriculture.

16. The implicit import-price elasticities are quite reasonable. In the case of dairy and meat, the base is very low, so the absolute change in imports is small.

17. A nominal payment of $11.9 billion times the change in the GNP deflator (1.154). See Economic Report of the President, 1987, p. 157.

18. See Economic Report of the President, 1987.

19. The real interest rate is defined as the annual prime rate minus the actual inflation rate. It jumped up in 1981 from 2 to about 8 percent and declined slowly thereafter.

20. See de Melo and Robinson (1985) and Dervis, de Melo, and Robinson (1981) for discussions of these effects in empirical models.

REFERENCES

Adelman, Irma and Sherman Robinson. "The Application of General Equilibrium Models to Analyze U.S. Agriculture." American Journal of Agricultural Economics, No. 5, 68(1986):1196-1207.

______________________________ (1988). "Macroeconomic Adjustment and Income Distribution: Alternative Models Applied to Two Economies." Journal of Development Economics, forthcoming.

Chambers, Robert G. and Richard E. Just. "Effects of Exchange Rates on U.S. Agriculture: A Dynamic Analysis." American Journal of Agricultural Economics 63(1982):249-265.

Condon, Timothy, Sherman Robinson, and Shujiro Urata. "Coping with a Foreign Exchange Crisis: A General Equilibrium Model of Alternative Adjustment Mechanisms." Mathematical Programming Study, No. 23, (1985):75-94.

Dervis, Kemal, Jaime de Melo, and Sherman Robinson. "A General Equilibrium Analysis of Foreign Exchange Shortages in a Developing Country." Economic Journal No. 364, 91(1981):891-906.

____ (1982). General Equilibrium Models for Development Policy. Cambridge: Cambridge University Press.

Dixon, Peter B. and Brian R. Parmenter (1986). "Medium-Run Forecasts for the Australian Economy Using the ORANI Model." Institute of Applied Economic and Social Research, Melbourne University.

Freebairn, John W., Gordon C. Rausser, and Harry de Gorter (1983). "Monetary Policy and U.S. Agriculture." Department of Agricultural and Resource Economics, Working Paper No. 266, University of California, Berkeley.

Henry, Mark and Gerald Schluter. "Measuring Backward and Forward Linkages in the U.S. Food and Fiber System." Agricultural Economics Research, No. 4 37(Fall 1985):33-39.

Hertel, Thomas W. and Marison E. Tsigas (1987). "Tax Policy and U.S. Agriculture: A General Equilibrium Analysis." Agricultural Economics Department, Purdue University, draft manuscript.

de Melo, Jaime and Sherman Robinson. "Trade Policy and Resource Allocation in the Presence of Product Differentiation." Review of Economics and Statistics, No. 2, 63(May 1981):169-177.

______________________________ (1985). "Product Differentiation and Trade Dependence of the Domestic Price System in Computable General Equilibrium Trade Models." In T. Peeters, P. Praet, and P. Reding, eds., International Trade and Exchange Rates in the Late Eighties, Amsterdam: North-Holland Publishing Co.

______________________________ (1986). "The Treatment of Foreign Trade in Computable General Equilibrium Models of Small Economies." Mimeo, Development Research Department, World Bank (June).

Robinson, Sherman (1986). "Multisectoral Models of Developing Countries: A Survey." Working Paper No. 401, Department of Agricultural and Resource Economics, University of California, Berkeley.

Robinson, Sherman and David W. Roland-Holst (1987). "Modelling Structural Adjustment in the U.S. Economy: Macroeconomics in a Social Accounting Framework." Working Paper No. 440, Department of Agricultural and Resource Economics, University of California, Berkeley.
Schuh, G. Edward. "The Exchange Rate and U.S. Agriculture." American Journal of Agricultural Economics 56(1974):1-13.
Shoven, John B. and John Whalley. "Applied General-Equilibrium Models of Taxation and International Trade." Journal of Economic Literature, No. 3, 22(September 1984):1007-1051.
U.S. Department of Commerce (1984). The Detailed Input-Output Structure of the U.S. Economy, 1977. U.S. Department of Commerce, Bureau of Economic Analysis.
U.S. Government (1987). Economic Report of the President, 1987.

Chapter 6

Christine Bolling

Price and Exchange Rate Transmission Revisited: The Latin America Case[1]

INTRODUCTION

In early discussions of the 1985 U.S. farm bill, two issues often came up. The first was concerned with the most likely price response in other countries to a change in the United States export price. If we define the U.S. Gulf port price as the world price, this issue could be addressed by simply comparing changes in other countries prices to a U.S. price change. The second issue concerned supply response to changing world prices. This can be addressed by studying changes in a country's domestic price to a changes in world prices through a price transmission equation and in turn the impact on a country's supply of a commodity.

Price transmission appears in the international trade literature as the bridge between the world price and a country's internal price. In the literature, Tweeten; Johnson; and Bredahl, Meyers and Collins account for price transmission through the use of an elasticity, e_{pi}, or the percentage change in a country's price resulting from a percentage change in the world price.

Tweeten and Johnson assumed that prices were transmitted perfectly, that is $e_{pi} = 1$. Bredahl, Meyers and Collins assigned 0 (no price transmission) or 1 (perfect price transmission) values to the price transmission elasticity, taking into account that most countries insulated domestic prices from world prices using policy instruments. In a previous study (1980), I demonstrated that there is a wide range of nominal farm price responses to any change in the world price, with exchange rates taken into consideration. By focusing on price transmission, we can isolate the effects of changes in world commodity prices and exchange rates on nominal domestic farm price, or alternatively, the price changes can be examined in real terms.

The specification of the price transmission equation begins with an identity, known as the price linkage equation, which links the domestic price of a commodity to the world price:

$$P_i = eP_w(1 + t) \qquad (1)$$

where P_i is the domestic (nominal) price in country i, P_w is the world price of the commodity, e is the exchange rate expressed in units of domestic currency per unit of a key currency (e.g., the U.S. dollar), and t are the transfer costs (transportation, tariffs, etc.). Assuming transfer costs remain constant, the price linkage equation (1) can be written in terms of percentage changes:

$$\frac{dP_i}{P_i} = \frac{de}{e} + \frac{dP_w}{P_w} \qquad (2)$$

Equation (2) says that the percentage change in country i's price is equal to the percentage change in the exchange rate plus the percentage change in the world price. To test the response of country i's price to changes in the exchange rate and changes in the world price, the following equation (expressed in logarithms) can be estimated:

$$\ln P_i = b_0 + b_1 \ln e + b_2 \ln P_w + U \qquad (3)$$

where U is the error term accounting for all other changes in P_i not accounted for by e and P_w.

In the free market case, where there is perfect transmission of prices and exchange rates, b_1 and b_2 would equal 1 in value. Less than perfect transmission would be indicated by values of b_1 and b_2 of less than 1. Extending the definition of the price-linkage equation to account for the effects of inflation, we have:

$$\frac{P_i}{D_i} = \frac{eD_w P_w (1 + t)}{D_i D_w} \tag{4}$$

where D_i is the domestic inflation index in country i, and D_w is the inflation index associated with the country whose price is used as the world price. Defining the real domestic price in country i as

$$P_i^* = \frac{P_i}{D_i}, \tag{5}$$

the real exchange rate as

$$e^* = \frac{eD_w}{D_i}, \tag{6}$$

and the real world price as

$$P_w^* = \frac{P_w}{D_w} \tag{7}$$

we can then write the price-linkage equation (in real terms) as

$$P_i^* = e^* P_w^* \,(1 + t). \tag{8}$$

Again assuming transfer costs remain constant, the price-linkage equation (8), expressed in real terms, can be written in terms of percentage changes:

$$\frac{dP_i^*}{P_i^*} = \frac{de^*}{e^*} + \frac{dP_w^*}{P_w^*} \tag{9}$$

which says that the percentage change in the real domestic price in country i is equal to the percentage change in the real exchange rate plus the percentage change in the real world price. Expressing equation (9) in logarithms for estimation purposes yields

$$\ln P_i^* = b_0 + b_1 \ln e^* + b_2 \ln P_w^* + U. \qquad (10)$$

In the case of free trade and assuming the law of one price holds, b_1 and b_2 should equal 1, that is, real changes in the exchange rate and real changes in world prices will be mirrored in real changes in domestic prices. If the law of one price does not hold or if a country insulates its market from world price movements by various policy instruments, then b_1 and b_2 will differ from 1 since real exchange rates and world price movements would not be completely transmitted to the domestic market.

COMMODITY PRICE BEHAVIOR

Comparing the internal farm prices of various Latin American countries for wheat, corn and soybeans with world prices[2] results in a wide range of responses to world prices. World prices for major agricultural commodities have displayed cyclical movements since the early 1970s, rising during the years 1970-1974 and 1977-1980, and falling during the years 1974-1977 and 1980-1984. Figure 6.1 shows how farm prices in U.S. dollars for wheat in selected Latin American countries moved with respect to the U.S. Gulf port price for wheat. Brazil, Bolivia, Colombia, Paraguay and Peru historically have had higher farm prices than the U.S. Gulf port price, while Argentina, Mexico, and Uruguay have had lower prices. All responded in some degree to the sharp world price increases of 1972-1974. Many countries, fearing continued shortages in the world market, increased farm prices to stimulate production, even if this was done with approximately a one year time lag. Only a few responded to the world price dip in 1977.

A more detailed look at how prices for wheat, corn, and soybeans in selected Latin American countries responded to world price changes indicates some consistent patterns (Table 6.1). Table 6.1 reports the number of years in which Latin American countries raised prices when world prices increased and likewise lowered prices when world prices declined. Except for Guatemala and Nicaragua in the case of corn, commodity prices in domestic currencies usually follow a world price increase. In contrast (with the exception of Paraguay in the case of corn and soybeans), prices in domestic currency did not usually follow a world price decline. The principal explanation for this type of price behavior is attributed to the existence of severe inflation in most Latin American countries during the period 1966-1984 which forced most to raise their commodity prices in spite of world price declines. Also, countries whose currencies were pegged to the U.S. dollar were more likely to follow a decline in world prices.

When converted to U.S. dollars by the official exchange rates, farm prices in Latin American countries were less likely to follow a world price increase but more likely to follow a world price decrease compared to when prices were expressed in domestic currencies. This price behavior reflects the impact of exchange rate adjustments on commodity prices, indicating that currency devaluations relative to the U.S. dollar in most Latin American countries have kept farm prices competitive with world prices. Countries whose farm prices (in U.S. dollars) were most likely to move in the same direction as world commodity prices (as evidenced by the larger diagonal numbers in Table 6.1) were: Argentina, Mexico and Uruguay in the case of wheat; Argentina, Chile, Mexico, Paraguay and Uruguay in the case of corn; and Argentina, Brazil, Colombia, Paraguay and Uruguay in the case of soybeans. The greater price adjustment in the case of corn and soybeans compared to wheat is consistent with the greater market orientation of government policies for corn and soybeans in most Latin American countries.

Table 6.1 shows that in most countries, prices in the domestic currency did not decline with falling world prices, primarily because inflation was severe in most Latin American countries. Countries which had their currencies pegged to the dollar were more apt to change their farm prices with a world price downturn. When internal farm prices were expressed in U.S. dollars, more countries had dollar price declines when the world dollar price declined, but several countries increased their farm prices with world price upturns. Mexico on wheat and corn, Brazil on soybeans, Paraguay on corn and soybeans, Uruguay on wheat and corn (as evidenced by the larger diagonal numbers in the table), were the most likely to change the direction of their prices with changing world prices.

Price policies, for these basic commodities, have sometimes shifted over the years with changing market conditions. Mexico, for example, maintained an artificially high producer price for wheat during the early 1960s; it then switched to prices that were below the Gulf port price in the late 1960s when surpluses developed. After a peak above the world price in 1981, domestic farm prices again fell below the world price in the early 1980s.

The extent of price transmission has been complicated by the shift in exchange rates relative to the U.S. dollar for many countries. There are a few countries like Guatemala, Honduras, Dominican Republic, Paraguay, and El Salvador whose currencies are officially pegged to the dollar. From 1966 to 1986, Nicaraguan currency fell in value between 20 to 100 percent (Table 6.2). Its currency has been pegged to the U.S. dollar for some time. Eleven countries, including Argentina, Brazil, Colombia, and Peru had exchange rates that fell in value by more than 100 percent vis-a-vis the dollar over the same period.

Latin American farm prices for wheat, as well as the world price, are all nominally at a higher level than in the pre-1972 era. In a few countries, internal farm prices (in terms of dollars) maintained roughly a constant percentage margin above (or below) world prices, or rose in proportion to international prices. For others, the relationship was inconsistent because of differences in inflation rates and exchange rate changes. Internal market conditions and domestic policies also caused this relationship to break down.

Wheat

Brazil and Colombia have traditionally priced their wheat well above the world market in order to protect their producers. Peru, Chile, Paraguay, and Guatemala also have farm prices for wheat that are well above the world price. In contrast, to accommodate consumer needs, Mexico maintained farm prices that were below the world price from 1972 to 1984 except in 1981. Argentina had farm prices below the world price except for the 1980-1981 period, which is consistent with its role as the region's major wheat exporter.

Corn

Most of Latin America had farm level corn prices that were generally above the world market price--Colombia, Bolivia, Peru,

Chile, Mexico, Guatemala, Nicaragua, Costa Rica, and El Salvador (Figure 6.2). Like many of these countries, Venezuela's corn is supported at a high level, and most is used for food. U.S. corn exports, in contrast, are almost totally used for feed. In several countries, corn is raised by Indians under primitive conditions, and prices are kept high to maintain income in isolated areas.

Soybeans

The major Latin American soybean exporters--Argentina, Brazil, Paraguay, and Bolivia have farm-level prices well below the U.S. Gulf port price, while Colombia, with its protected soybean market, has high farm prices relative to the U.S. Gulf port price (Figure 6.3). Mexico, in most years, also has had farm prices for soybeans set above the world market.

COUNTRY PRICE POLICIES

An overview of the countries' agricultural policies in Latin America provides insight into the level of market interference by government and the resulting price and exchange rate transmission that results. During 1966-1980, most countries in Latin America maintained control of marketing and foreign trade of basic commodities through a parastatal organization under a fixed exchange rate regime, so changes in world prices did not necessarily translate into changes in farm prices in Latin America. As countries added on layers of policies that interfered with the free market, less price and exchange rate transmission occurred.

Most Latin American countries have modified their level of control as world economic conditions have changed. In this section, the discussion is limited to agricultural policy developed during the 1980s. Some countries introduced extreme policy changes during 1966-1980 that affected both price and exchange rate transmission elasticities and so changes in their pre-1980 policies are also discussed where appropriate.

Argentina

Argentine export policy of multiple exchange rates and export taxes has been the central focus of Argentine farm policy administered through the Junta Nacional de Granos. Domestic grain and oilseed prices have been subject to downward pressure resulting from policies designed to dampen inflation, raise government revenues, and redistribute income to the industrial sector and urban population.

Argentina's farm policy is much different than the policy of most other Latin nations whose goals generally are to protect farm income or stabilize prices. Export taxes and differential exchange rates, however, have been traditional sources of revenue for the Argentine government. Through the years these policy measures have been applied to grains, oilseeds, oilseed meals, and vegetable oils.

Argentina has experienced inflation and subsequent devaluations for some time. Since the government had a practice of preventing the official exchange rate from adjusting to fully account for domestic inflation, each devaluation ostensibly provided the opportunity for windfall profits, hence, the need for export taxes. Export taxes hit their peak during the second Peronist era when they ranged from 30 to 50 percent. During the early eighties, export taxes were considerably lower.

In mid-1982, when Argentina was in the midst of massive devaluation, export taxes gained new importance. At that time the export tax was 25 percent on grains and soybeans. Grains had traditionally been taxed and soybeans had only recently come on the export scene. Policies covering oilseed products changed from rebates (reembolso), used to encourage exports of high value oilseed products, to a differential export tax of 10 percent that was also in the same spirit.

Brazil

Brazil supports producer prices of wheat, corn, soybeans, cotton, and rice. The Brazilian government has traditionally priced its wheat well above the world market to protect its producers. It has also provided subsidies for inputs such as fertilizers. Retail prices of wheat products have been subsidized in order to benefit consumers. This consumer subsidy was eliminated in early 1988. Brazil has maintained differential export taxes for soybean products.

On the import side, it has also used import licensing and foreign exchange rationing, especially for wheat and corn. In the early 1980s Brazil had a crawling-peg exchange rate with mini devaluations. (From March to September 1986 the Cruzado was fixed vis-a-vis the dollar.)

Bolivia

Bolivia's agriculture has been influenced by fixing producer prices for basic commodities and controlling domestic processing and marketing at the consumer level. Some of these price controls are beginning to be dismantled. The domestic market is isolated from the international market through the use of export quotas. In recent years, Bolivia has had numerous controls on producer prices and marketing margins as a means of keeping consumer prices low.

Peru

Peru has had a succession of parastatal organizations.[3] Through these agencies, and their monopoly control on imports and exports, the Peruvian government insulated essential commodities from the sharp price fluctuations in the world market. But these same organizations kept producer prices at low levels as one means of subsidizing consumers in addition to the program of direct food subsidies to low income families. The Public Enterprise of Agricultural Services (EPSA) was set up in 1969 to handle the wholesale marketing of cotton, corn, wheat, milk and meat. In 1974, the government established the Marketing Board for Fish Meal (EPCHAP) which dealt with import and export of all agricultural commodities, except for edible oils which were the responsibility of Marketing Board for Inputs (ENCI). In 1980, EPSA transferred its responsibility for marketing all its commodities except rice to ENCI. In the same year, rice marketing was transferred to Marketing Board for Rice (ECASA), giving it a trading monopoly on rice. Through the 1970s the Regulating Commission of Agricultural Products of Metropolitan Lima (JURPAL) was instituted to fix marketing margins and wholesale prices for foodstuffs, but its functions were later transferred to the Ministry of Commerce.

During the 1980-1985 period, the Peruvian government attempted to eliminate price controls and food subsidies. In 1983 milk, bread, pasta, and wheat prices were freed from price controls. Wheat prices, in particular, were adjusted to more closely reflect world market prices. Since then the increase in costs has been passed on to consumers. To try to maintain some limit on the price increase, there has been a price control on "pan corriente" (ordinary bread). Since the new government was elected in 1985, the food price policy has shifted back to one of price controls. The government reinstated ENCI as the sole importer of wheat, feedgrains, oilseed products, and fertilizers in 1985. Peru has annual global quotas on major bulk agricultural commodities that are approved by the Council of Ministers. Peru also has some tariffs. The Peruvian government raised tariffs on corn and sorghum from 3 to 15 percent in October 1982, but dropped then to 10 percent in January 1983. The government established a 15 percent ad valorem tariff on all products in 1984, but established duty exonerations on wheat, non-fat dry milk, butter oil, whole dry milk, and rice. In January 1985, imports of a number of luxury commodities, such as wine and cigarettes, were suspended. The Garcia government put a tax on the c.i.f. value of all imports effective January 1986. The import tax is to be used to finance rural development, through guaranteed support prices.

Peru aggressively devalued its currency during most of the 1980s. The nominal rate of depreciation of the sol against the U.S. dollar was 65 percent in 1982, 133 percent in 1983, 112 percent in 1984 and 216 percent in 1985.

Venezuela

In Venezuela, farm prices for corn, rice, and cotton have been generally supported at a very high level, and the internal market has been protected by import quotas. Tariff barriers for grains are not significant. Feedgrains and foodgrains are exempt from tariffs. There is a 3.5 percent import tax on oilseed products. The binding constraints on imports are import permits and foreign exchange approvals through the Ministry of Agriculture and Foreign Exchange Commission which are required before foreign exchange is made available by the Central Bank. Foreign exchange is rationed, amounting to U.S. $8 billion in 1985, because of the Government's desire to build up its foreign exchange reserves.

As a guideline to importers, the Ministry of Agriculture establishes an "Import Program" each year. Import quotas are allocated to the different firms in the industry on the basis of their plant capacity. The Ministry tries to force purchase and consumption of domestic crops before imports are allowed. In the case of oilseeds, a new government policy for 1986 required that the import quota allocations of oilmeal and vegetable oil be tied to crushing capacity, a measure intended to encourage mills to expand their crushing capacity, rather than to continue imports of soybean oil and meal. Venezuela has successfully improved its external situation by progressively devaluing its currency, the bolivar, within a multiple rate system. The "primary" exchange rate, which applies to most imports, is 7.5 bolivares per dollar. But from 1982 to 1985, basic food goods (wheat, sorghum, corn, soybeans) were still imported at the 4.3 bolivares rate, an implicit import subsidy. The favorable rate of exchange for imports of essential food products was changed to 7.5 bolivares in 1986. Capital transfers, travel, and some imports are invoiced at the free market rate of about 20 bolivares per dollar. In 1985, 10 percent of imports occurred at the free market rate, 10 percent at a 4.3 rate, and the rest at a 7.5 rate. The government of Venezuela receives almost all of the dollars generated by the export sector because of the nationalized oil sector and so can set exchange rate where it wants, within the limits of world oil prices and its own foreign reserves.

Mexico

Mexico, during the 1972-1980 period, had wheat prices that were well below world price levels. The Mexican government has used the National Popular Subsistence Company (CONASUPO), its state trading organization, as the instrument to separate the domestic market from the world market. Since CONASUPO was the sole importer of these basic agricultural commodities, it controlled imports and established the wedge between the higher import price and the lower domestic producer and consumer prices. Mexico currently guarantees farm prices for wheat, corn, cotton, soybeans and their products. In most years these are below the world market price. The government subsidizes basic foods through CONASUPO. The government also owns such diverse enterprises as feed and milk processing plants and grocery stores to maintain low meat and dairy prices to consumers.

Traditionally, Mexico has taken a protectionist stance in international trade, requiring import licensing for grains, oilseeds, and oilseed products. The Mexican government has only allowed private traders to import grain and oilseeds since 1985.

Ecuador

Ecuador has minimum prices for basic commodities that are maintained by the National Agency for Food Storage and Marketing (ENAC) and the National Agency for Essential Products (ENPROVIT). ENAC is responsible for the purchasing, storage, handling, and resale of these commodities, while ENPROVIT sells food at the retail level. The government also sets maximum prices for flour and rice. Responsibility for the agricultural price policy is delegated to the Ministry of Agriculture. The minimum price for soybean meal is set by the Ministry of Agriculture and Ministry of Industry, Commerce and Economic Integration.

In foreign trade, wheat, corn, soybeans, and rice are subject to import licensing, while soybean and vegetable oil are subject to import quotas. The imports of rice, corn, and soybeans are at prices below official domestic prices. Wheat imports are subsidized when the import price is greater than the reference price, and taxed when they are less. Imports of agricultural inputs, like fertilizer and other agrochemicals, are subject to permit requirements that are seldom enforced. Exchange rates are also used to control trade. Ecuador has a "crawling peg" exchange rate and differential exchange rates. Agricultural commodities and inputs have been imported at the Central Bank's market intervention rates since June 1985.

Chile

Chile has generally had a free market regime except for wheat and oilseeds. The Chilean government in 1983 reinstated the price support program that had been eliminated in 1979. The Chileans set their farm price in accordance with the world price, but maintain an import price band whereby wheat and oilseed imports are allowed if the import price falls within the price band. The price band mechanism, which amounts to a variable levy, was created to protect domestic production. If the price of imported wheat is below the price band, even after payment of the 30 percent ad valorem tariff, a surcharge (sobretasa) is added to bring the landed price up to at least the floor of the price band. Since the general tariff level dropped in 1986 from 35 percent to 30 percent, the government increased the surcharge to keep the band in effect. The floor price is recalculated weekly, based on a moving average of the price of wheat in the last 60 months. Currently, the effect of the price band is to tax imports and raise the cost of wheat to consumers. When world wheat prices increase, the implied producer subsidy increases. A similar price band exists for the oilseeds sector. Chile devalued its currency continuously between 1982 and 1986 and maintained a unified exchange rate for all imports.

Colombia

Colombia has supported farm prices for wheat and soybeans at a relatively high level through the Agricultural Marketing Institute (IDEMA), its central marketing agency. Colombia also has producer price support programs for corn and rice. Colombia has subsidized agricultural inputs, particularly fertilizers. The Colombian government has also subsidized exports of cotton and rice through its indirect tax rebate certificate (CERT) system for exports, during 1966-1980. Colombia has a history of price controls on basic commodities, although in the past they were not necessarily effective. During 1966-1980 IDEMA was also the sole importer of wheat. A prohibitively high tariff was charged to keep private firms from importing wheat. IDEMA was permitted to collect the equivalent of the import tariff to carry on its own marketing activities. The Ministry of Foreign Trade issues licenses for wheat and corn imports, and foreign exchange is rationed for wheat, corn, sorghum, and soybeans and their products through an import quota system. The quota level is set every trimester, and the monthly budget for all imports in 1985 was limited to $250 million. The allocations of agricultural import quotas are based on the past import record of agroindustrial firms. There are also non-tariff barriers in Colombia, such as the high ($40/ton) fee for use of ports. In their exchange rate policy, the monetary authorities of Colombia have opted for a gradual devaluation in the 1980s, rather than a one-time shock.

EMPIRICAL ESTIMATES OF TRANSMISSION ELASTICITIES

This section tests the relationship between the U.S. Gulf port price and the domestic producer price for corn, wheat, and soybeans in selected Latin American countries. Both the nominal price transmission and exchange rate transmission elasticities were estimated using the following model:

$$\log P_i = b_0 + b_1 \log e + b_2 \log P_w + U \qquad (11)$$

where P_i is the internal farm price of the commodity considered and is expressed in the currency of a specific country, P_w is the U.S. Gulf port price, and e is the exchange rate between the U.S. dollar and the country's domestic currency.[4] The model provides a measure of the statistical relationships between countries' supply prices, the exchange rate, and the U.S. export price for a specific commodity. This model is tested with alternative assumptions that: 1) internal farm prices changed in the same year as the U.S. Gulf port price and 2) internal farm prices changed with a one year time lag, because policy constraints may prevent immediate responses to world price changes.

The latter results are reported when they provided better statistical fit. The resulting coefficients on the exchange rate and the Gulf port price were tested to determine if they were significantly different from 1.[5] The real price and exchange rate transmissions were also estimated using the following equation:

$$\log P_i^* = b_0 + b_1 \log e^* + b_2 \log P_w^* + U \qquad (12)$$

where P_i is the internal farm price of the commodity considered and is expressed in the currency of the specific country and is deflated by the country's GDP deflator, P_w^* is the world price defined here as the U.S. Gulf port price, deflated by the U.S. GDP deflator, and e* is the exchange rate between the U.S. dollar and the country's currency deflated by their respective GDP deflators. The resulting coefficients were tested to determine if they were significantly different from 1.[6] The time period has been extended from the earlier Collins study (1980) to include the years 1966-1984. Results were compared with the earlier study that included data through 1976. This study employs the Food and Agriculture Organization (FAO) data base of internal farm prices for wheat, corn, and soybeans in selected Latin American countries. The U.S. Gulf port price for these commodities is used as a measure of world price (USDA, ESCS).

Price and exchange rate transmission elasticities, even in nominal terms, can range between zero and 1.[7] It can be misleading in policy models to assume a priori that elasticities are exactly zero or 1. Tables 6.3 through 6.5 summarize the results of the regression analyses.

Wheat Results

For wheat, Chile and Uruguay exhibit nearly perfect price and exchange rate transmission when their farm price and exchange rate are expressed in nominal terms (Table 6.3). Perfect price transmission demonstrates that these countries' farm prices changed in accordance with the world price and the exchange rate. Paraguay has also had statistically perfect price transmission, but has kept the same exchange rate with the dollar since 1960. Since 1961, Argentina, the region's largest producer and exporter, has transmitted exchange rate changes but the export tax system has prevented price transmission for wheat.

If internal farm prices for wheat are deflated by the country's GDP deflator and are compared with the deflated U.S. Gulf port wheat price, Uruguay comes closer than other countries in Latin America to having real internal farm prices that move with the real U.S. Gulf port wheat price. Chile appears to adjust its real wheat price to the world price of the previous year, which reflects the difference in its growing season compared with the U.S. Chile has also had problems of adjusting farm prices in a climate of rapid inflation in some years.

Corn Results

Brazil and Colombia are the only corn-producing countries whose nominal price and exchange rate transmission elasticities are close to 1.0, and, therefore, whose farm prices are closely related to the world price and exchange rate movements (Table 6.4). Chile has attempted to keep its farm prices up with world prices, but adjustments to rapid inflation in some years has caused its nominal prices to increase faster than the world price.

If real internal farm prices (nominal farm price divided by the GDP deflator) for corn are compared with real Gulf port price, no country has perfect price transmission.

Soybean Results

Argentina, Brazil, Colombia, Mexico, and Paraguay all exhibit perfect price transmission for soybeans (Table 6.5), when prices and exchange rates are expressed in nominal terms. Brazil is the principal U.S. competitor in the international soybean market and Argentina has rapidly expanded soybean area and exports. Bolivia is about the only Latin American producer that has not kept its nominal internal prices in line with the U.S. Gulf port price, even though production has increased rapidly and Bolivia has become an exporter of soybean oil and meal. The nominal exchange rate has also been transmitted to the farm level soybean price in Argentina, Brazil, Colombia and Mexico. In real terms, Argentina, Brazil, and Paraguay come the closest to perfect price transmission among Latin American countries.

When countries make abrupt changes or insulate their markets from world price fluctuations through their policies, the resulting price and exchange rate elasticities will be different between time periods, as evidenced by the results of the earlier Collins study (1980) compared with this study. The results for Mexico's wheat and corn, Argentina's corn and soybeans, and Colombia's soybeans are cases in point.

CONCLUSIONS

It is well known that in formulating U.S. policy it is important to understand what other countries' price policy reactions are likely to be. We have the tool for analyzing these reactions and a firm historical perspective using the concept of price transmission. This paper has concentrated on producer prices and the supply side, even though there is an equally relevant consumer side with consumer price reactions.

A ranking of nominal and real price transmissions (Table 6.6 and 6.7) gives an indication of how much nominal farm prices change with world market prices. Overall we see that there is a mixed reaction among Latin American countries, as measured by price transmission. The degree of reaction depends on the magnitude of the distortion caused by their price and non-price policies and exchange rate policies. We have been able to isolate the relative influence of world price changes and exchange rate changes on nominal farm price changes. In the case of real prices, we have been able to isolate the relative influence of real world price changes and real exchange rate changes.

While the price elasticities have been estimated here to be constant as a consequence of the double log functional form, there is no reason to expect these elasticities to be the same in the eighties as they were during the sixties and seventies. They depend too much on the vagaries of politics in each country.

Figure 6.1. Wheat: Internal Farm Price Compared with World Price

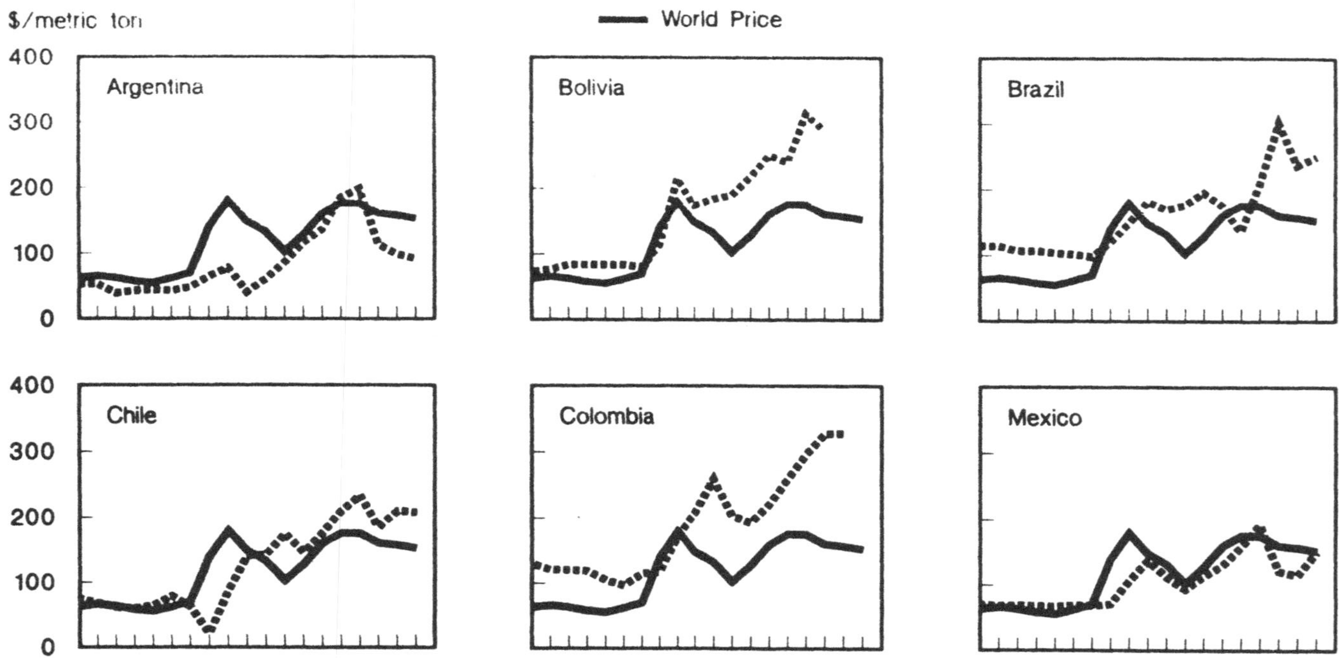

Figure 6.1. Continued.

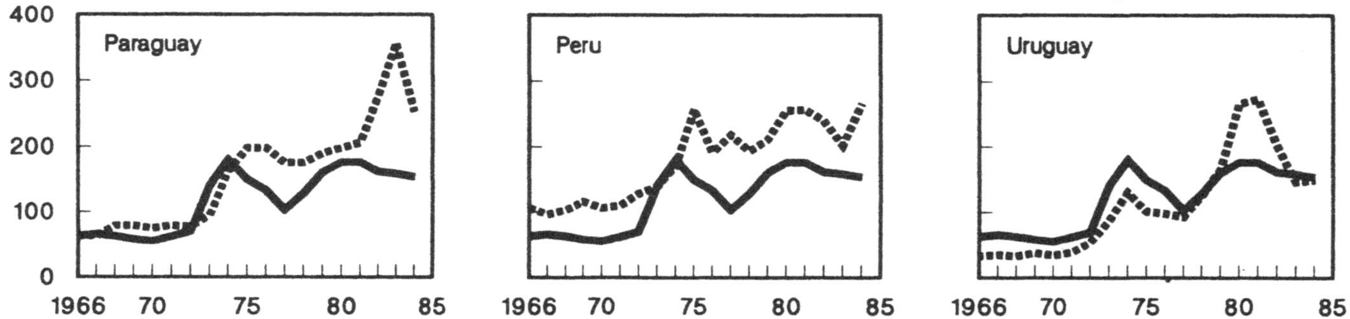

Figure 6.2. Corn: Internal Farm Price Compared with World Price

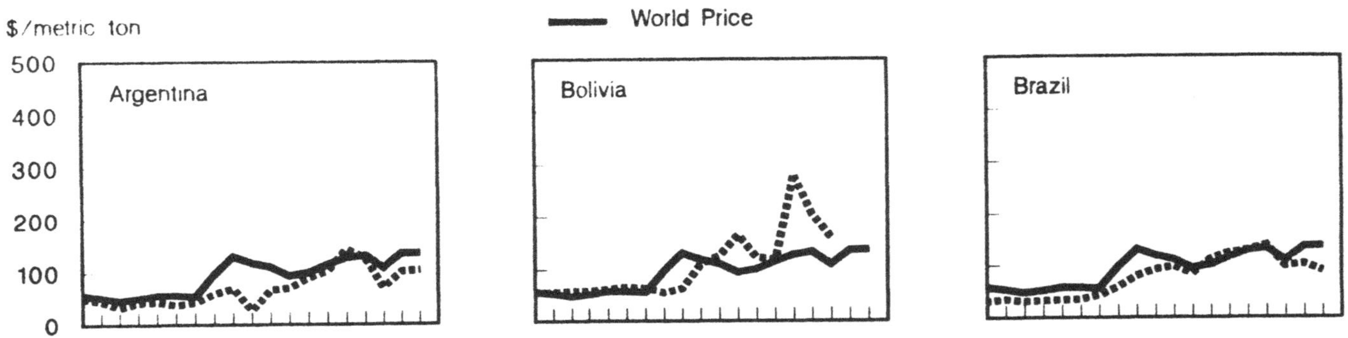

Figure 6.2. Continued.

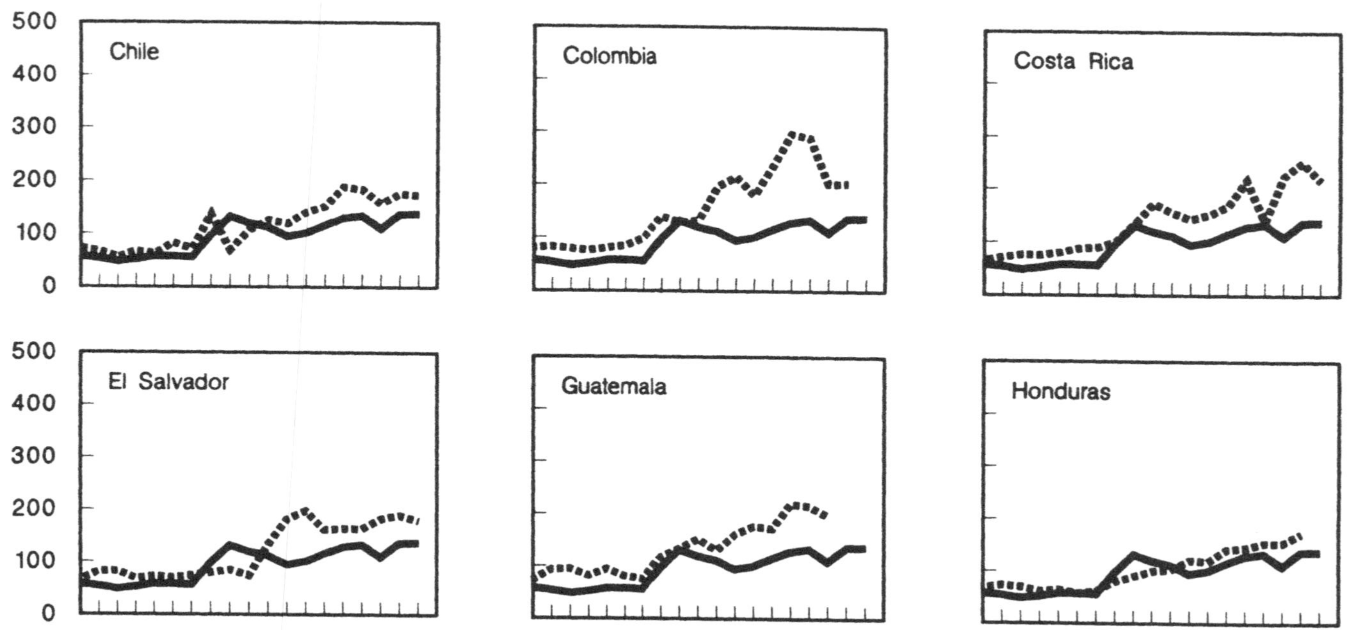

Figure 6.2. Continued.

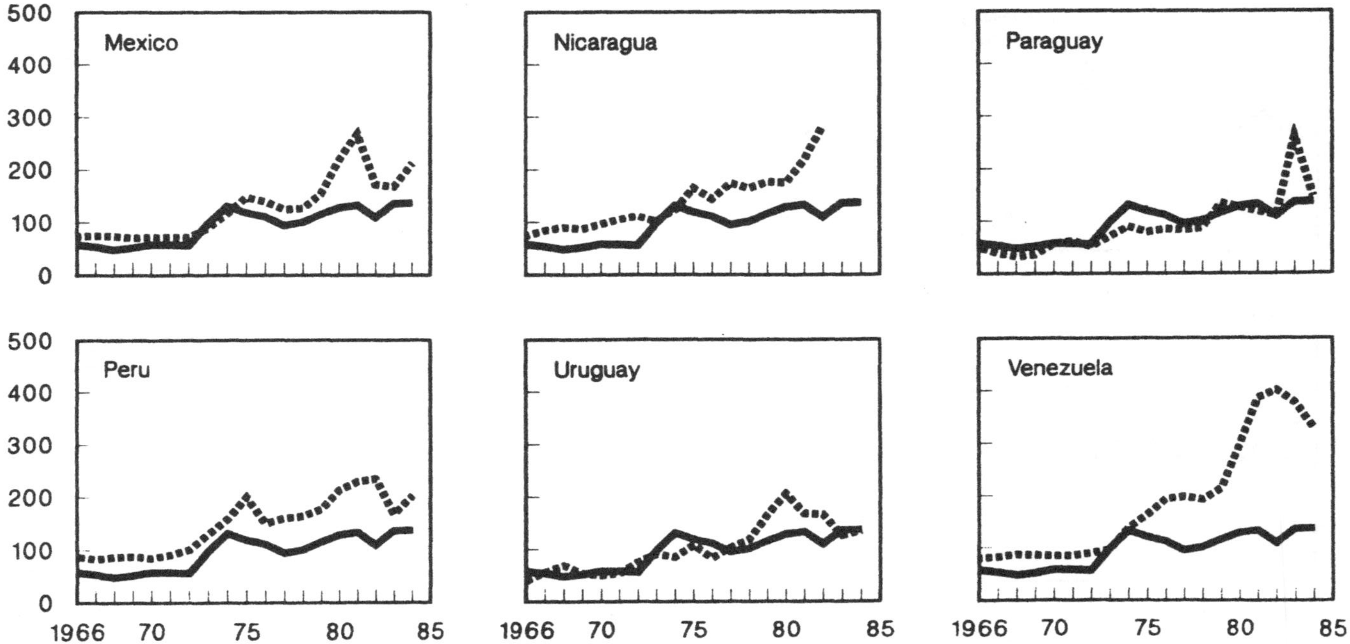

Figure 6.3. Soybeans: Internal Farm Price Compared with World Price

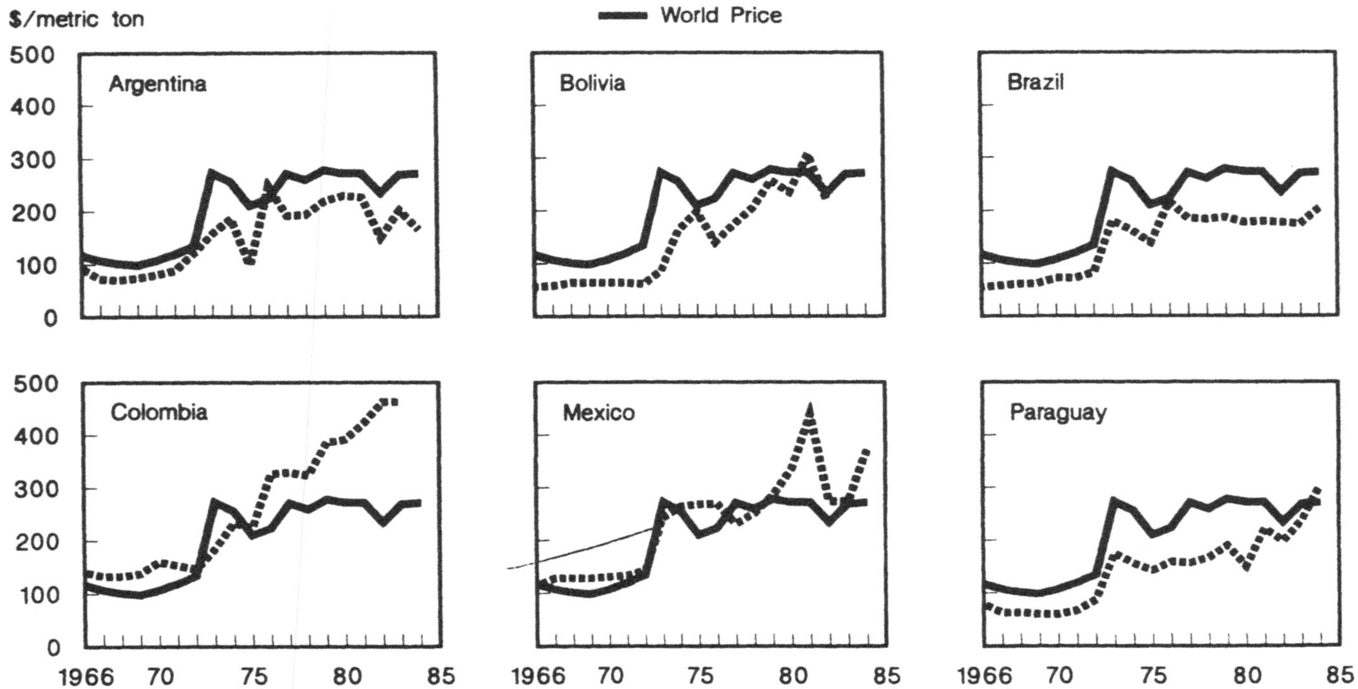

Source: Calculated for FAO Farm Price Data Base.
International Financial Statistics (6, 7).

Table 6.1. Direction of Change in Prices from Previous Year by Country, 1966-1984

	Wheat		Corn		Soybeans	
	World price increase	World price decline	World price increase	World price decline	World price increase	World price decline
Argentina						
Currency price increase	9	9	11	7	12	6
Currency price decline	0	0	0	0	0	0
Dollar price increase	8	4	9	3	8	3
Dollar price decline	1	5	2	4	4	3
Bolivia*						
Currency price increase	9	8	10	7	9	6
Currency price decline	0	1	1	0	1	0
Dollar price increase	7	7	5	5	8	4
Dollar price decline	2	2	4	2	2	2
Brazil						
Currency price increase	9	9	11	7	11	6
Currency price decline	0	0	0	0	1	0
Dollar price increase	5	5	10	4	8	3
Dollar price decline	4	4	1	3	4	3
Chile						
Currency price increase	9	9	11	7	–	–
Currency price decline	0	0	0	0	–	–
Dollar price increase	5	5	7	3	–	–
Dollar price decline	4	4	4	4	–	–
Colombia						
Currency price increase	8	8	10	6	11	6
Currency price decline	1	1	0	1	0	0
Dollar price increase	6	6	7	5	7	3
Dollar price decline	3	3	3	2	4	3
Costa Rica						
Currency price increase	–	–	10	5	–	–
Currency price decline	–	–	1	2	–	–
Dollar price increase	–	–	8	5	–	–
Dollar price decline	–	–	3	2	–	–
El Salvador						
Currency price increase	–	–	7	6	–	–
Currency price decline	–	–	4	1	–	–
Dollar price increase	–	–	7	6	–	–
Dollar price decline	–	–	4	1	–	–
Guatemala						
Currency price increase	6	6	4	4	–	–
Currency price decline	3	3	7	3	–	–
Dollar price increase	6	6	4	4	–	–
Dollar price decline	3	3	7	3	–	–

Table 6.1. Continued.

	Wheat		Corn		Soybeans	
	World price increase	World price decline	World price increase	World price decline	World price increase	World price decline
Honduras						
Currency price increase	–	–	7	6	–	–
Currency price decline	–	–	4	1	–	–
Dollar price increase	–	–	7	6	–	–
Dollar price decline	–	–	4	1	–	–
Mexico						
Currency price increase	7	7	10	6	12	6
Currency price decline	2	2	1	1	0	0
Dollar price increase	7	3	9	2	9	5
Dollar price decline	2	6	2	5	3	1
Nicaragua						
Currency price increase	–	–	5	4	–	–
Currency price decline	–	–	6	3	–	–
Dollar price increase	–	–	5	4	–	–
Dollar price decline	–	–	6	3	–	–
Paraguay						
Currency price increase	8	7	9	1	10	2
Currency price decline	1	2	2	6	2	4
Dollar price increase	8	6	8	1	9	2
Dollar price decline	1	3	3	6	3	4
Peru						
Currency price increase	7	8	10	7	–	–
Currency price decline	2	1	1	0	–	–
Dollar price increase	7	5	9	5	–	–
Dollar price decline	2	4	2	2	–	–
Uruguay						
Currency price increase	8	9	9	7	–	–
Currency price decline	1	0	2	0	–	–
Dollar price increase	9	2	5	1	–	–
Dollar price decline	0	7	6	6	–	–
Venezuela						
Currency price increase	–	–	7	7	–	–
Currency price decline	–	–	4	0	–	–
Dollar price increase	–	–	5	7	–	–
Dollar price decline	–	–	6	0	–	–

Source: Calculated from FAO and IMF data.
*No change in exchange rate.

Table 6.2. Currency Devaluations, 1966-1986

0-20 percent	20-100 percent	More than 100 percent
Dominican Republic	Nicaragua	Argentina
El Salvador		Bolivia
Guatemala		Brazil
Honduras		Chile
Paraguay		Colombia
		Costa Rica
		Ecuador
		Mexico
		Peru
		Uruguay
		Venezuela

Source: IMF Data.

Table 6.3. Wheat: Estimates of Price and Exchange Rate Transmission Elasticities for Selected Latin American Countries

Country	Estimation period	Nominal price transmission country currency	Previous study	Nominal exchange rate transmission	Previous study	Real price transmission country currency	Real exchange rate transmission
Argentina	1966-80	1.012(x)	1.014(x)	1.152	0.361		
	1966-84	0.550		1.051(x)		-0.270	0.101
Bolivia	1966-80	0.196	2.310				
	1966-84	0.780		1.370		0.366	-0.029
Brazil	1966-80	0.169	0.199	1.228(x)	0.932(x)		
	1966-84	0.304		1.104(x)		0.123	1.139
Chile	1966-80	1.264(x)		.975(x)			
	1966-84	0.397		1.076		-0.250	1.050
Colombia	1966-80	0.667		1.155(x)			
	1966-83	0.303		1.486		0.177	-0.173
Mexico	1966-80	0.668(x)	0.541	1.290(x)	*		
	1966-84	0.567		1.081(x)		0.084	0.183
Paraguay	1966-80	1.006(x)			*		
	1966-84	1.040(x)		1.622		0.440	0.271
Peru	1966-80	0.529		1.157			
	1966-84	0.599		1.064(x)		0.179	0.364
Uruguay	1966-80	1.054(x)	1.349(x)	1.107(x)	.955(x)		
	1966-84	1.120(x)		1.086(x)		0.663	-0.073

(x) Indicates that coefficient is not significantly different from 1, for perfect price transmission.
*No change in exchange rates.

Table 6.4. Corn: Estimates of Price and Exchange Rate Transmission Elasticities for Selected Latin American Countries

Country	Estimation period	Nominal price transmission country currency	Previous study	Nominal exchange rate transmission	Previous study	Real price transmission country currency	Real exchange rate transmission
Argentina	1966-80	1.373(x)	1.115(x)	1.071(x)	0.453		
	1966-84	0.421		1.057(x)		-0.159	0.353
Bolivia	1966-80	1.386	0.250	0.777	0.238		
	1966-84	-0.257		1.189(x)		-0.266	0.044
Brazil	1966-80	1.038(x)	1.101(x)	1.074(x)	0.782(x)		
	1966-84	1.312(x)		1.001(x)		0.501	0.006
Chile	1966-80	0.511	0.694(x)	0.909(x)	1.001(x)		
	1966-84	0.252	1.067(x)				
Colombia	1966-80	0.734(x)		1.121(x)			
	1966-83	0.620(x)		1.414		0.423	0.021
Costa Rica	1966-80	0.712		1.006(x)			
	1966-84	0.818		1.178(x)		0.268	0.815
Dominican Republic	1966-80	0.700(x)		*			
	1966-84	1.040(x)		1.622		0.197	-0.185
Ecuador	1966-80	0.670		1.369		0.593	0.769
	1966-84						
El Salvador	1966-80						
	1966-84	0.839		*		-0.140	0.419
Guatemala	1966-80	0.695	0.465	*	*		
	1966-84	0.872(x)		*		0.339	0.227
Honduras	1966-80	0.701		*		0.216	-0.031
	1966-84						
Mexico	1966-80	0.886(x)	0.455	1.905(x)	*		
	1966-84	0.887(x)		1.111(x)		0.061	0.048
Nicaragua	1966-80						
	1966-84					0.760	-0.100
Paraguay	1966-80	1.033(x)		*			
	1966-84	1.172(x)		1.532(x)		0.248	-0.169
Peru	1966-80	0.765		1.117			
	1966-84	0.807(x)		1.037(x)		0.353	0.454
Uruguay	1966-80	0.411	0.470	1.116(x)	1.102(x)		
	1966-84	0.229		1.175(x)		0.071	0.482
Venezuela	1966-80	0.625	0.661	7.621	0.303		
	1966-84	1.282		1.657		-0.249	-0.151

(x) Indicates that coefficient is not significantly different from 1, for perfect price transmission.
*No change in exchange rates.

Table 6.5. Soybeans: Estimates of Price and Exchange Rate Transmission Elasticities for Selected Latin American Countries

Country	Estimation period	Nominal price transmission country currency	Previous study	Nominal exchange rate transmission	Previous study	Real price transmission country currency	Real exchange rate transmission
Argentina	1966-80	0.821(x)	0.429	1.015(x)	0.491		
	1966-84	0.959(x)		1.001(x)		0.723	0.739
Bolivia	1966-80	0.013		0.189			
	1966-84	0.651		0.641		0.623	0.181
Brazil	1966-80	0.886(x)	1.105(x)	1.125(x)	0.734(x)		
	1966-84	1.141(x)		1.012(x)		0.860	0.584
Colombia	1966-80	0.777(x)	0.380	1.043(x)	1.077(x)	0.184	0.068
	1966-83						
Mexico	1966-80	0.913(x)	0.926(x)	*	*		
	1966-84	0.868(x)	1.060(x)			0.413	0.209
Paraguay	1966-80	1.077(x)		*			
	1966-84	1.151(x)		1.739		0.210	0.314

(x) Indicates that coefficient is not significantly different from 1, for perfect price transmission.

Table 6.6. Ranking of Nominal Price Transmission, 1966-1984

.80 - 1.00	.60 - .80	0 - .60
Wheat		
Chile	Argentina	Bolivia
Paraguay	Colombia	Brazil
Uruguay	Peru	
	Mexico	
Corn		
Bolivia	Argentina	Uruguay
Brazil	Colombia	Venezuela
Mexico	Costa Rica	
Paraguay	Dominican Republic	
Chile	Guatemala	
	Honduras	
	Peru	
Soybeans		
Argentina	Colombia	Bolivia
Brazil		
Mexico		
Paraguay		

Table 6.7. Ranking of Real Price Transmission, 1966-1984

.80 - 1.00	.60 - .80	0 - .60
Wheat		
Uruguay	Chile	Argentina
		Bolivia
		Brazil
		Colombia
		Mexico
		Paraguay
		Peru
Corn		
	Chile	Argentina
		Brazil
		Colombia
		Costa Rica
		Dominican Republic
		Guatemala
		Honduras
		Mexico
		Paraguay
		Peru
		Uruguay
		Venezuela
Soybeans		
Argentina		Bolivia
Brazil		Colombia
Paraguay		Mexico

NOTES

1. The author expresses her appreciation to C. Carter, C. S. Kim, L. Deaton, B. Krissoff, W. Gardiner and R. King for their excellent reviews.
2. Defined as the U.S. Gulf port price for U.S. No. 1 Hard Red Winter Wheat, U.S. Gulf port price for U.S. No. 2 Yellow Corn and U.S. Gulf port price for U.S. No. 2 Yellow Soybeans.
3. Parastatal organizations are quasi government organizations which are usually attached to the Ministry of Agriculture and charged with regulating prices, domestic markets, and trade of particular commodities. They usually derive their income from user fees, and thus have a vested interest in their business that goes beyond the government that they serve.
4. The exchange rate used was the official exchange rate (IMF).
5. The composite hypothesis is $b_0 = 0$, $b_1 = 1$, and $b_2 = 1$.
6. The composite hypothesis is $b_0 = 0$, $b_1 = 1$, and $b_2 = 1$.
7. Chambers disputes this bounding in (3).

REFERENCES

Argentina, Bolsa de Cereales de Buenos Aires. Bolsa de Cereals. Buenos Aires, selected issues.

Bredahl, Marury E., William H. Meyers, and Keith J. Collins. "The Elasticity of Foreign Demand for U.S. Agricultural Products: The Importance of the Price Transmission Elasticity," American Journal of Agricultural Economics 61(No. 1, February 1979).

Chambers, Robert C. and Richard E. Just. "A Critique of Exchange Rate Treatments in Agricultural Trade Modeling," American Journal of Agricultural Economics (May 1979):249-257.

Collins, H. Christine. "Price and Exchange Rate Transmission," Agricultural Economics Research 32(No. 4, October 1980).

________________. "Price Transmission between the U.S. Gulf Port and Foreign Farm Markets," IED Staff Report, Economics, Statistics, and Cooperative Services, U.S. Department of Agriculture, 1978.

Food and Agriculture Organization. "Farm Price Data Base" (unpublished). Available on USDA TDAM computer system and computer tape from FAO.

International Monetary Fund. International Financial Statistics. Selected issues.

Jabara, Cathy, L. and Nancy E. Schwartz. "Flexible Exchange rates and Commodity Price Exchanges, The Case of Japan," Economic Research Service Staff Report No.AGES 860624, Washington, D. C., August 1986.

Johnson, Paul R. "The Elasticity of Foreign Demand for U.S. Agricultural Products," American Journal of Agricultural Economics 59(1977):735-736.

McCalla, Alex and Timothy E. Josling. Agricultural Policies and World Markets, MacMillian Publishing Company, New York, 1985.

Tweeten, Luther. "The Demand for United States Form Output," Food Research Institute Study 7, 1967.

U.S. Department of Agriculture, Economics, Statistics, and Cooperative Service. "ESCS Price Data Base" (unpublished). Available on USDA TDAM system.

Chapter 7

David Blandford

Market Share Models and the Elasticity of Demand for U.S. Agricultural Exports

INTRODUCTION

In many international trade models it is assumed that a given product is relatively homogeneous and that imports from different countries are highly substitutable. The perfect substitution assumption is frequently used in deriving direct and indirect estimates of the excess demand equation facing a single country, and in multi-country spatial and non-spatial price equilibrium models of international trade.[1] The assumption implies that each supplier faces the same underlying demand elasticity in a given market, although the excess elasticity facing a supplying country may be different depending on market share and the supply elasticities of competitors. If the perfect substitution assumption does not hold, this has important implications. The price elasticity of demand facing each supplier will be lower. This will reduce the effect that any price discounting might have on import volumes and may lead to a fall in total supplier revenues if such discounting is used. Alternatively, it may imply that exporters could capture monopoly rents if they were to restrict supplies to price inelastic markets.

There are two major reasons why the imports from different suppliers may not be perfect substitutes. Differences in the technical characteristics of products may affect demand at the intermediate or final levels. For example, the protein content of wheat influences its milling and baking properties. Imported wheats of differing protein contents may therefore be imperfect substitutes for one another. Market rigidities may also influence the demand for imports from competing suppliers. Importing firms may have commercial relationships which limit their willingness to substitute among the products of competing suppliers. Traditions of language and custom or imperfect information may mean that imports from different suppliers appear to be imperfect substitutes. In this case, products which are technically indistinguishable are distinguished by place of production.

The relaxation of the perfect substitution assumption has led to the development of a whole class of empirical trade models. Some of the multi-country versions of these models are discussed by Goddard. In this chapter, the focus will be on those models which can be used to analyze changes in imports by origin in a single importing country. Since the emphasis is on the United States, the focus will be narrowed further to the relative responsiveness of imports from the United States versus those from the rest of the world. Although this approach does not permit analysis of the competitive relationship between the United States and particular export competitors, it does permit an evaluation of the overall responsiveness of U.S. market share in a number of importing countries.

The analysis employs three types of models: (1) the Markov model; (2) the elasticity of substitution equation; and (3) the market share equation. These models are applied to data on the quantity of imports of wheat and corn from the United States and other suppliers in China (PRC), the European Community (EC), Egypt, Japan, Mexico, South Korea, Taiwan and the USSR. The data are quarterly for the years 1970 through 1983 and were obtained from Figueroa who provides a thorough description of these data and their source. Both quarterly and annual models are estimated, where appropriate. The choice of data period is an important issue in the empirical analysis of international trade flows and therefore receives attention.

Tables 7.1 and 7.2 summarize imports of wheat and corn from the United States in the markets studied for selected years, and the corresponding market share that these imports represented. In many cases, imports and U.S. market share have been highly variable. Some markets, such as China, have been affected by substantial changes in domestic agricultural and trade policies. Others have been influenced by changes in U.S. policy. The USSR is the most dramatic example with the U.S. export moratoria of the mid-1970s and the grain embargo of 1980-1981. In many of the markets analyzed, imports in total and from the United States have displayed high variability. Explaining changes in these imports is a considerable challenge for any economic model.

THE MARKOV MODEL

The Markov model has been used by a number of authors to analyze changes in import market shares (Dent; Atkin and Blandford). The model assumes that the average import share of the jth exporter in a particular market is a random variable following a first order Markov process

$$m_{j,t} = \sum_{i=1}^{r} m_{i,t-1} p_{ij} + v_{j,t} \tag{1}$$

where $m_{j,t}$ is the share of the jth exporter in the market in t; $m_{i,t-1}$ is the market share of the ith exporter in t-1; p_{ij} is the probability that imports will switch from i to j; $v_{j,t}$ is a statistically independent error term; and r is the number of exporters. The model implies that the probability of an outcome (current market share) depends only on the outcome in the previous period, and that this probability is constant for all time periods.

The p_{ij}'s are called transitional probabilities. They can be arranged as an rxr matrix whose elements have the following properties

$$0 \leq p_{ij} \leq 1 \tag{2}$$

$$\sum_{j} p_{ij} = 1 \tag{3}$$

In order to impose these conditions, the transitional probability matrix may be estimated through quadratic programming in which the sum of the squared residuals derived from equation (1) are minimized subject to equations (2) and (3). A recursive method is used to improve the accuracy of the solution. An extensive discussion of the Markov approach, the properties of alternative estimators, and the algorithm used in this chapter are contained in Lee et al.

One of the problems of applying the Markov approach is that the sample size which can be handled computationally is limited. The available algorithm will only accommodate a maximum of six market shares and 30 observations. Furthermore, because of the assumptions which underlie the model, the derivation of as table solution is difficult for data which display substantial variability from period to period. Both of these characteristics limit the application of the model to quarterly data. As a result, the model was estimated annually. The results obtained are summarized in Tables 7.3 and 7.4.

These tables contain estimates of the three relevant market share probabilities for the United States: the repeat purchase, gain and loss probabilities. The mean square error (MSE) of the predicted market shares (U.S. and rest of world) are included in each table as an indicator of the "goodness of fit" of the model.

The repeat purchase probability is an indicator of loyalty on the part of an importer. Other things being equal, a high repeat purchase probability will mean that the United States will tend to maintain its market share through time; a low probability will mean that US market share will tend to be more variable from period to period. The actual variability of market share, and its trend, will also be dependent on the relative size of the gain and loss probabilities.

The gain probability indicates the extent to which the United States will tend to capture market share from its competitors (the rest of the world). The loss probability indicates the degree to which the United States will tend to lose market share to its competitors. From a U.S. perspective, a desirable combination of circumstances is one in which a high repeat purchase probability is combined with high gain and low loss probabilities. This combination of probabilities will reflect as table but growing market share.

Despite the simplicity of the Markov model, the MSE indicates that in the majority of cases the model fits the data reasonably well. The fit is best for those countries such as the EC and Japan, which tend to have the greatest stability in their importing behavior. The fit is poorer for more variable purchasers such as the PRC and Mexico.

The results suggest that the repeat purchase probability for corn is high in most markets, ranging from .78 for the USSR to .97 in the EC. The repeat purchase probabilities for wheat are much more variable, ranging from zero in the EC to .94 in Egypt and Mexico. The gain and loss probabilities are highly variable. The gain probabilities range between zero for corn in the EC and wheat in Mexico to a high of .80 for wheat in Korea. The loss probabilities range between .03 for wheat in Egypt to 1 for wheat in the EC.

It is difficult to derive an overall impression of the market characteristics implied by the Markov model from the estimates in the two tables. Table 7.5 contains rankings of the markets based on two alternative criteria. For the first criterion the loss probability was subtracted from the sum of the repeat purchase and gain probabilities. This provides a rough indicator of how favorable the conditions were for the United States in each market under the assumption that the stability of market share is important. As indicated above, the combination of large repeat purchase and gain probabilities and a small loss probability could be viewed to be desirable, although this may not represent the best possible combination of circumstances. With a high preference for growth and low preference for stability, a high gain/low loss combination alone might be considered preferable. This is reflected by criterion 2 in which the loss probability is subtracted from the gain probability.

Both of the criteria produce rankings which are broadly similar. From a U.S. perspective, the South Korean wheat market appears to have desirable characteristics; the EC wheat market has clearly undesirable features. The South Korean market for both wheat and corn are both highly ranked using the two criteria. The Chinese (PRC) wheat market, the Soviet wheat market, both rank low. There are many cases in which the market characteristics for one product in a given country appear to be favorable from a U.S. point of view, while the characteristics for the other product are unfavorable. For example, the Japanese and Mexican corn markets both appear to have desirable characteristics, but their wheat markets are much less desirable.

Finally, the Markov model can be used to provide some insight into a controversial issue -- the effects of the Soviet Grain embargo on Soviet import behavior. In Tables 7.3 and 7.4 two estimates of the Markov probabilities are provided for the U.S.S.R. The first of these refers to the entire data period (1970-1983); the second omits the post-embargo years since 1980. For both corn and wheat the model suggests a worsening of the characteristics of the Soviet market from a U.S. perspective. The estimates for the pre-embargo period suggest a higher repeat purchase probability, higher gain probability and lower loss probability than when the post-embargo years are included. The deterioration is particularly noticeable for wheat.

THE ELASTICITY OF SUBSTITUTION MODEL

The Markov model provides some insight into changes in U.S. market share for the commodities and countries analyzed. It suggests which markets have had favorable characteristics in terms of changes in market share. However, because it is a probabilistic model it provides little insight into the responsiveness of purchases from the United States to economic variables, particularly price. One model which has been used to analyze this question is the import elasticity of substitution model.

In the context of imports, the elasticity of substitution is defined simply as the proportionate change in the relative quantities imported from two competing suppliers divided by the proportionate change in their import prices. It is typically estimated using annual data from the following equation

$$\log\left(\frac{m_{1t}}{m_{2t}}\right) = b_0 + b_1 \log\left(\frac{p_{1t}}{p_{2t}}\right) + v_t \qquad (4)$$

where m denotes imports, p is price and the subscripts 1 and 2 relate to the competing suppliers. A negative sign is expected on the price variable, in that an increase in the price of imports from supplier 1 should result in a decline in the purchases from that supplier relative to those from the competing supplier 2. In the current case supplier 1 is the United States and supplier 2 is the rest of the world.

The results of applying this model to annual data for the eight markets analyzed in this chapter are summarized in Tables 7.6 and 7.7. The dependent variable is the log of the ratio of imports from the United States to those from all other suppliers. In those cases where imports from the United States or competing suppliers were zero, an arbitrary small quantity (10 tons) was used in order to permit the use of logarithms. This procedure was also followed in the estimation of the market share equation. The independent variable is the log of the ratio of the price of imports from the United States to a weighted average of the prices from other competitors. The weights used are current period imports. The U.S. and competitor prices are in domestic currency units. For further details on the basic data see Figueroa.

An examination of the results in Tables 7.6 and 7.7 shows that the model performs extremely poorly. Even though there are only 14 observations, the model explains an extremely small proportion of the total variation in the import ratio. The Durbin-Watson "t" statistic suggested that many of the OLS equations were highly autocorrelated. These equations were re-estimated using the Cochrane-Orcutt procedure. The estimated value of the autocorrelation coefficient is given in the table, and its order. All estimates were derived using the micro-TSP econometric package. Most of the estimated substitution elasticities are of the anticipated negative sign. The only positive coefficients are for corn in China and wheat in Mexico. However, only three of the price ratio coefficients are statistically significant from zero at the 5 percent confidence level. These are corn in Mexico, and wheat in Egypt and Japan. The price coefficient on corn in Japan is significant at the 10 percent level.

The statistical quality of these results is similar to that in previous agricultural applications. For example, Capel and Rigaux in a study on Canadian and U.S. wheat reported estimates of variable statistical quality. In fact, they found a number of statistically significant substitution elasticities which were positively signed. Positive values are difficult to justify either theoretically or intuitively.

It is possible that the poor quality of these results is due to the use of annual data. If markets are reasonably competitive, it is likely that changes in the ratio of competing import prices will not persist for a long period of time. Competitive pressures would be likely to return the price ratio to "normal" levels. However, sustained changes in the ratio of prices could exist over shorter time periods. In order to evaluate whether this is the case, the model was re-estimated using quarterly data. The only additional variables included were a set of quarterly dummy variables to account for any regular variation in quarterly import patterns.

The results of applying the model to quarterly data are summarized in Tables 7.8 and 7.9. Problems with autocorrelation remain, and it was necessary to use the Cochrane-Orcutt procedure in more than half of the equations. Nevertheless, the number of correctly signed variables which are statistically significant at the 5 percent confidence level increases to 8, compared to 3 in the annual model. These results, if accepted, would suggest that a significant substitution elasticity exists in half of the markets analyzed and that these elasticities are substantially in excess of one in value, with the exception of corn in the EC.

As indicated by Leamer and Stern, the concept of the elasticity of substitution has its theoretical foundation with respect to movements along a single consumer indifference curve. In order to obtain a valid estimate of the substitution elasticity from equation (4) it is necessary that the sum of the direct and cross price elasticities of the two commodities should be equal, and that the income and other price elasticities of demand for the two commodities also should be equal. In order to test that these conditions are satisfied, Leamer and Stern suggest that the following unrestricted model be estimated

$$\log\left(\frac{m_1}{m_2}\right) = b_0 + b_1 \log p_1 + b_2 \log p_2 + b_3 \log y + b_4 \log p_n + v_t \qquad (5)$$

where y denotes income and p_n denotes the prices of other commodities.

There are statistical problems with the estimation of this model. It is likely that p_1 and p_2 will be highly correlated with each another, and possibly with p_n. If an undeflated total income variable is used, multicollinearity is likely to be worse. The precision of the estimators will be reduced, as will the power of significance tests. It is possible to reduce, but not entirely eliminate this problem, by using real per capita income (GDP) in the equation. The price of other goods is reflected by the consumer price index. Both of these variables are unavailable for the PRC. The correlation between the U.S. price and the competitor price was generally high, frequently around .99. The results obtained from the unrestricted elasticity of substitution model are contained in Tables 7.10 and 7.11.

A "t" test was used to evaluate whether there is a statistically significant difference in the absolute value of the p_1 and p_2 coefficients. In no case could the null hypotheses of no difference be rejected at the 5 percent confidence level. The own price elasticities are sufficiently similar to satisfy the assumptions underlying the elasticity of substitution model. A second requirement is that both the coefficients on the price of other commodities (the CPI) and income should not be significantly different from zero. This condition is violated for one or both of the coefficients in 5 of the corn equations and 1 wheat equation. As a result, the elasticity of substitution estimates given in Tables 7.8 and 7.9 must be rejected as theoretically invalid for corn in the European Community and South Korea, and wheat in the Soviet Union. Only the estimates for corn in Egypt (-12.5), and Mexico (-10.4) and the estimates for wheat in Mexico (-27.6), South Korea (-18.9), and Taiwan (-10.4) may be retained.[2]

THE MARKET SHARE MODEL

As this analysis has demonstrated, there are problems in the application and interpretation of the elasticity of substitution equation in the current case. A practical, but less theoretically elegant, alternative is provided by the market share equation. This may be specified as

$$M_t = b_0 + b_1 M_{t-1} + b_2 \left(\frac{p_{1t}}{p_{2t}} \right) + v_t \tag{6}$$

where M denotes U.S. market share. The equation is typically estimated in a linear or log-linear form (Sirhan and Johnson).

One of the advantages of the market share equation is that it allows for lags in response to price changes, even though this is achieved through the restrictive assumption of a geometrically declining lag structure. Given the rigidities and inflexibilities that often exist in international markets, lags in import response are likely to occur.

Tables 7.12 and 7.13 contain a summary of the results of applying this model to annual data. The linear or log-linear equation is reported depending on which functional form explains the higher proportion of the total variability in the dependent variable. As in the case of the elasticity of substitution model, the statistical quality of the estimates is poor. Only 9 of the 32 lagged share or price coefficients are statistically different from zero at the 10 percent confidence level or above. Those markets which appear to be responsive to relative prices are the same markets which displayed a statistically significant response to price in the elasticity of substitution model.

The estimation of the model using quarterly data produces a substantial improvement in the quality of the results (Tables 7.14 and 7.15). The number of coefficients which are statistically significant at the 10 percent level or above increases to 20 from the 9 obtained with annual data; correctly signed and statistically significant price coefficients are found for 9 of the 16 markets. Particularly for wheat there appears to be evidence of widespread response to changes in relative prices, although the effect of the price change on imports is generally delayed. The short and long-run effects of a sustained 1 percent decrease in the ratio of U.S. prices to those of competitors are summarized in Table 7.16. These effects are calculated using the sample mean price ratio for linear equations and the sample mean import share for the log-linear equations. Most of the values appear to be reasonable with the possible exception of corn in Egypt. The estimate suggests that U.S. market share in the Egyptian market is extremely sensitive to price.

As for the Markov model, the market share equation can be used to examine the impact of the Soviet grain embargo on the determinants of U.S. market share. From Table 7.14 it may be seen that there is an increase in the value of the price coefficient for corn in the post-embargo period, but neither coefficient is significantly different from zero at the 5 percent level. In the wheat case, there is a similar increase in price response and both coefficients are statistically significant. However, the difference in the value of the coefficients is not statistically significant. The Markov model results given above suggest that there was a change in the determination of market shares. The current results suggest that this was not related to a change in the response to relative prices.

It is possible to use the information derived from the market share equation to derive a ranking of the markets from a U.S. perspective. Markets can be divided into three categories: (1) those that are responsive to changes in relative prices; (2) those that are not price responsive but where U.S. market share in the current quarter show a systematic relationship to the share in the previous quarter; and (3) those for which U.S. market share appears to be a random variable. It is assumed that price responsive markets are the most desirable, since a reduction in the U.S. price relative to those of its competitors would imply that the United States would gain market share. Note, however, that this would not necessarily imply that total revenue would increase. The impact of a price reduction on total imports and revenues cannot be evaluated from the market share model.

Markets which are not price responsive but display continuity in market share from period to period are assumed to be the next most desirable set. If no gain in market share can be achieved from a reduction in price, at least the United States can be guaranteed some continuity of market share. This second set of markets is ranked on the basis of the value of the lagged share coefficient. An increasing value of the coefficient is similar in interpretation intuitively to high repeat purchase probabilities in the Markov model.

Table 7.17 contains a ranking of the markets based on these definitions and assumptions. The EC is again low on the list, as it was in the earlier ranking obtained from the Markov model results. In general, developing countries are highly ranked, while the centrally planned markets are generally in the lower half of the order. Japan is the lowest ranked of the price-sensitive markets, but given the criteria adopted is a more consistently favorable market than its Asian counterparts.

CONCLUSIONS

The principal focus of this chapter has been the responsiveness of U.S. market share in a number of important markets for wheat and corn. A secondary focus has been the comparative performance of the elasticity of substitution and market share models, and the effects of using quarterly rather than annual data.

The application of the Markov model to annual data provides some pointers to the stability of U.S. market share and the competitive position of the United States in the markets analyzed. However, it does not provide any information on price responsiveness. The annual elasticity of substitution and market share models are weak statistically and tend to suggest that little response to changes in relative prices exists in the markets studied. When applied to quarterly data this misconception is dispelled. The results derived from the elasticity of substitution equation must be discounted in many cases because the assumptions underlying the model do not hold. However, the market share equation suggests that the price of U.S. imports relative to those of competitors is a significant determinant of market share in over half of the markets examined.

The results of the analysis, although limited in their scope by the simplicity of the models employed, have some important implications. First, the choice of data periodicity may influence significantly conclusions about the responsiveness of trade volumes to price. In competitive markets, much of the price response which exists may not be captured in annual models. Second, the evidence of differential response to price across markets is important from a policy perspective. Because of the considerable variation across markets, the effects of a general reduction in U.S. export prices could differ significantly. Assuming that competitors do not retaliate, targeted subsidies could have an effect on market share in some markets but little effect in others. The results obtained in this study provide some information relevant to these important issues. However, a definitive assessment of the impact of changes in U.S. domestic and trade policies on the volume of U.S. exports would require more complicated multi-equation models.

Table 7.1. Imports of Corn from the United States and U.S. Market Share

Country	1970	1975	1980	1983
Imports from the U.S.	---------------- Thousand metric tons ----------------			
China (PRC)	0	0	1,667	1,356
EC	6,709	12,457	10,468	3,729
Egypt	19	509	983	1,380
Japan	4,193	5,369	11,823	11,979
Mexico	492	1,408	4,850	4,459
South Korea	51	354	2,313	4,073
Taiwan	0	607	1,997	3,266
USSR	0	3,173	4,228	2,371
Share of the above	---------------------- Percent ----------------------			
markets in U.S. total	79.6	71.4	60.8	75.7
U.S. market share	---------------------- Percent ----------------------			
China (PRC)	0.0	0.0	92.4	86.4
EC	59.7	69.9	67.0	34.6
Egypt	20.7	93.6	73.8	76.7
Japan	88.5	99.8	92.1	80.1
Mexico	63.2	51.7	100.0	89.0
South Korea	100.0	66.7	93.1	92.5
Taiwan	0.0	43.6	76.7	93.5
USSR	0.0	57.2	42.3	37.6

Table 7.2. Imports of Wheat from the United States and U.S. Market Share

Country	1970	1975	1980	1983
Imports from the U.S.	---------------- Thousand metric tons ----------------			
China (PRC)	0	0	6,105	2,458
EC	2,029	2,470	1,529	852
Egypt	0	1,038	1,686	3,860
Japan	2,760	2,919	3,331	3,465
Mexico	3	67	676	5
South Korea	1,508	1,590	1,945	1,884
Taiwan	544	410	551	621
USSR	0	4,084	1,770	4,936
Share of the above	------------------------ Percent ------------------------			
markets in U.S. total	35.4	39.3	47.4	43.6
U.S. market share	------------------------ Percent ------------------------			
China (PRC)	0.0	0.0	50.8	21.5
EC	18.1	21.4	14.4	9.9
Egypt	0.0	34.7	32.9	66.6
Japan	56.3	51.2	58.1	52.7
Mexico	100.0	75.3	66.1	1.2
South Korea	97.6	94.0	96.4	91.5
Taiwan	34.9	70.3	80.2	88.3
USSR	0.0	36.0	11.4	20.3

Table 7.3. Markov Model: Corn

Import market		------U.S. Probabilities------ Repeat purchase	Share gain	Share loss	MSE
China (PRC)*		0.818	0.506	0.182	0.106
EC		0.970	0.000	0.030	0.006
Egypt		0.843	0.683	0.157	0.010
Japan		0.935	0.521	0.065	0.005
Mexico		0.876	0.469	0.124	0.054
South Korea		0.897	0.611	0.103	0.014
Taiwan		0.964	0.156	0.036	0.020
USSR	1970-83	0.784	0.451	0.216	0.043
	1970-79	0.849	0.568	0.151	0.031

*Excludes 1976 and 1977 when there were no imports.

Table 7.4. Markov Model: Wheat

		------U.S. Probabilities------			
Import market		Repeat purchase	Share gain	Share loss	MSE
China (PRC)		0.722	0.098	0.278	0.028
EC		0.000	0.196	1.000	0.001
Egypt		0.941	0.088	0.059	0.015
Japan		0.556	0.554	0.444	0.002
Mexico*		0.940	0.000	0.060	0.052
South Korea		0.914	0.802	0.086	0.007
Taiwan		0.904	0.370	0.096	0.015
USSR	1970-83	0.275	0.269	0.725	0.022
	1970-79	0.402	0.273	0.598	0.027

*Excludes 1976 when there were no imports.

Table 7.5. Ranking of U.S. Markets Based Upon the Markov Model

	Criterion 1		Criterion 2
South Korea wheat	1.630	South Korea wheat	0.716
South Korea corn	1.405	Egypt corn	0.526
Japan corn	1.391	South Korea corn	0.508
Egypt corn	1.369	Japan corn	0.456
Mexico corn	1.221	Mexico corn	0.345
Taiwan wheat	1.178	China (PRC) corn	0.324
China (PRC) corn	1.142	Taiwan wheat	0.274
Taiwan corn	1.084	USSR corn 1970-83	0.235
USSR corn 1970-83	1.019	Taiwan corn	0.120
Egypt wheat	0.970	Japan wheat	0.110
EC corn	0.940	Egypt wheat	0.029
Mexico wheat	0.880	EC corn	-0.030
Japan wheat	0.666	Mexico wheat	-0.060
China (PRC) wheat	0.542	China (PRC) wheat	-0.180
USSR wheat 1970-83	-0.181	USSR wheat 1970-83	-0.456
EC wheat	-0.804	EC wheat	-0.804

Criterion 1 = repeat purchase + share gain - share loss probabilities.
Criterion 2 = share gain - share loss probabilities.

Table 7.6. Elasticity of Substitution Model: Corn, Annual Data

Import market	Constant	Price ratio (log)	AR	R2
China (PRC)	-3.352	30.106	0.64 (1)	0.00
	(-0.59)	(1.07)	(2.72)	
EC	-9.568	-0.493	0.99 (1)	0.30
	(-0.06)	(-0.86)	(3.23)	
Egypt	2.406	-0.692		0.00
	(1.21)	(-0.04)		
Japan	2.865	-10.321 **	0.52 (1)	0.22
	(3.38)	(-1.93)	(2.09)	
Mexico	-5.584	-44.867 *		0.26
	(-1.15)	(-2.06)		
South Korea	2.678	-0.185	0.62 (1)	0.01
	(0.71)	(-0.01)	(3.00)	
Taiwan	1.148	-2.612	0.31 (1)	0.01
	(0.72)	(-0.38)	(2.42)	
USSR	1.170	-1.889	0.15 (1)	0.00
	(2.16)	(-0.27)	(1.25)	

*Statistically significant at the 5% level or above.
**Statistically significant at the 10% level or above.
Value of the "t" statistic in parentheses.
AR denotes autocorrelation coefficient, order given in parentheses; for autocorrelated models, the R2 relates to the original OLS model.

Table 7.7. Elasticity of Substitution Model: Wheat, Annual Data

Import market	Constant	Price ratio (log)	AR	R2
China (PRC)	-4.980	-1.113		0.08
	(-3.32)	(-0.99)		
EC	-1.656	-0.036		0.00
	(-13.62)	(-0.06)		
Egypt	0.212	-43.469 *	0.619 (2)	0.34
	(0.10)	(-3.54)	(4.12)	
Japan	0.188	-4.389 *		0.39
	(4.96)	(-2.76)		
Mexico	1.158	11.821	0.707 (1)	0.00
	(0.15)	(0.30)	(2.30)	
South Korea	3.222	-6.450		0.04
	(3.56)	(-0.72)		
Taiwan	4.187	-10.873		0.19
	(2.32)	(-1.69)		
USSR	1.749	-2.931	0.668 (1)	0.09
	(0.10)	(-0.37)	(2.46)	

*Statistically significant at the 5% level or above.
Value of the "t" statistic in parentheses.
AR denotes autocorrelation coefficient, order given in parentheses.

Table 7.8. Elasticity of Substitution Model: Corn, Quarterly Data

Import market	Constant	Price ratio	AR (1)	R2
China (PRC)	-0.860	-3.351	0.489	0.04
	(-0.69)	(-1.01)	(3.84)	
EC	-0.151	-0.952 *	0.803	0.06
	(-0.79)	(-3.78)	(8.73)	
Egypt	5.545	-12.485 *		0.12
	(4.51)	(-2.15)		
Japan	1.551	1.094	0.624	0.08
	(1.78)	(0.35)	(5.37)	
Mexico	1.226	-10.388 *		0.20
	(1.07)	(-2.99)		
South Korea	4.513	-11.878 *	0.385	0.16
	(5.05)	(-3.40)	(2.99)	
Taiwan	2.233	-7.618	0.534	0.02
	(1.63)	(-1.52)	(4.57)	
USSR	0.825	-0.476	0.613	0.53
	(0.93)	(-0.20)	(5.865)	

(1) First order autocorrelation coefficient.
For equations corrected for autocorrelation the R2 relates to the uncorrected OLS equation.
*Statistically significant at the 5% confidence level.
Value of the "t" statistic in parentheses.
All equations estimated with quarterly dummy variables.

Table 7.9. Elasticity of Substitution Model: Wheat, Quarterly Data

Import market	Constant	Price ratio	AR (1)	R2
China (PRC)	-3.439 (-2.07)	-5.264 (-0.97)	0.811 (9.88)	0.17
EC	-0.828 (-12.11)	0.033 (0.15)		0.18
Egypt	0.549 (0.30)	-2.450 (-1.31)		0.92
Japan	0.029 (0.79)	-0.887 (-1.85)		0.14
Mexico	0.236 (0.19)	-27.591 * (-3.176)	0.399 (2.82)	0.20
South Korea	5.923 (6.76)	-18.866 * (-4.12)		0.29
Taiwan	4.376 (3.35)	-10.403 * (-2.51)		0.12
USSR	-0.741 (-0.72)	-14.723 * (-2.67)	0.607 (5.57)	0.53

(1) First order autocorrelation coefficient.
For equations corrected for autocorrelation the R2 relates to the uncorrected OLS equation.
*Statistically significant at the 5% confidence level.
Value of the "t" statistic in parentheses.
All equations estimated with quarterly dummy variables.

Table 7.10. Unrestricted Elasticity of Substitution Model: Corn, Quarterly Data

Import market	Constant	U.S. Price	ROW Price	CPI	Income	R2
China (PRC)	-9.212	2.709	0.563			0.05
	(-0.72)	(0.27)	(0.07)			
EC*	2.394	-0.77	-0.849	0.29	0.584	0.24
	(1.94)	(-1.78)	(-1.09)	(1.32)	(2.59)	
Egypt	2.675	-27.252	35.870	-2.552	0.139	0.18
	(0.32)	(-1.96)	(2.30)	(-0.50)	(0.02)	
Japan*	-18.037	10.550	-1.258	-4.432	-0.190	0.31
	(-0.81)	(1.22)	(-0.14)	(-2.99)	(-0.07)	
Mexico	-3.253	-21.614	17.405	2.460	2.709	0.35
	(-0.40)	(-2.51)	(2.11)	(0.94)	(0.80)	
South Korea*	38.309	-35.694	23.878	4.887	-0.561	0.31
	(3.01)	(-4.00)	(2.90)	(2.15)	(-0.19)	
Taiwan*	-31.816	-29.193	36.750	-6.880	14.105	0.63
	(-3.00)	(-3.56)	(3.76)	(-3.35)	(5.17)	
USSR*	369.855	-9.398	-8.911	-75.445	20.592	0.53
	(4.54)	(-1.58)	(1.42)	(-4.162)	(6.50)	

*Indicates that the restricted model would be rejected because of the significance of the coefficient on the cpi and/or income variable.

Table 7.11. Unrestricted Elasticity of Substitution Model: Wheat, Quarterly Data

Import market	Constant	U.S. Price	ROW Price	CPI	Income	R2
China (PRC)	-40.155	-37.734	52.262			0.42
	(-5.16)	(-1.94)	(18.68)			
EC	-1.798	-0.124	0.529	-0.035	0.139	0.23
	(-1.33)	(-0.21)	(0.455)	(-0.17)	(0.76)	
Egypt	-22.139	-24.416	30.440	0.594	1.573	0.61
	(-6.04)	(-1.94)	(2.66)	(0.25)	(0.54)	
Japan	-1.238	-1.502	1.632	0.026	0.075	0.19
	(-1.07)	(-1.29)	(1.42)	(0.38)	(0.46)	
Mexico	1.639	-28.213	31.355	-3.962	3.604	0.29
	(0.19)	(-1.26)	(1.47)	(-1.35)	(0.79)	
South Korea	-9.335	-41.859	46.058	-1.561	0.792	0.34
	(-1.00)	(-3.94)	(4.35)	(-0.78)	(0.28)	
Taiwan	-22.054	-17.352	23.555	1.747	-1.843	0.32
	(-2.55)	(-1.70)	(2.36)	(0.63)	(-0.47)	
USSR*	383.592	-18.698	21.681	-89.595	15.300	0.43
	(3.53)	(-1.22)	(1.42)	(-3.749)	(3.59)	

*Indicates that the restricted model would be rejected because of the significance of the coefficient on the cpi and/or income variable.

Table 7.12. Market Share Equation: Corn, Annual Data

Import market	Constant	Lagged share	Price	R2
China (PRC)	0.166 (0.139)	0.512** (1.99)	0.138 (0.13)	0.30
EC	-0.107 (-0.45)	1.001* (2.92)	0.123 (0.88)	0.49
Egypt	0.977 (1.74)	0.143 (0.82)	-0.258 (-0.56)	0.13
Japan	0.954 (2.81)	0.310 (1.12)	-0.344** (-1.79)	0.35
Mexico	2.326 (3.09)	0.239 (0.99)	-2.142* (-2.42)	0.42
South Korea	0.444 (0.77)	0.271 (0.88)	0.177 (0.30)	0.11
Taiwan	0.809 (0.90)	0.921* (3.83)	-0.579 (-0.72)	0.69
USSR	3.337 (0.48)	0.164 (1.31)	-2.332 (-0.35)	0.15

*Statistically significant at the 5% level or above.
**Statistically significant at the 10% level or above.
Value of the "t" statistic in parentheses.
All equations are linear.

Table 7.13. Market Share Equation: Wheat, Annual Data

Import market	Constant	Lagged share	Price	R2
China (PRC)	0.405 (1.87)	0.423 (1.74)	-0.283 (-1.42)	0.46
EC	0.210 (2.60)	-0.411 (-1.09)	0.027 (0.31)	0.11
Egypt (1)	-1.290 (-1.50)	0.537* (3.69)	-35.770* (-2.36)	0.77
Japan	1.825 (3.98)	-0.044 (-0.19)	-1.258* (-2.90)	0.46
Mexico	3.162 (1.94)	0.427 (1.22)	-2.986 (-1.67)	0.26
South Korea	0.783 (0.73)	0.089 (0.25)	0.036 (0.03)	0.01
Taiwan	1.444 (1.55)	0.624* (3.40)	-0.868 (-1.22)	0.55
USSR	0.853 (1.34)	-0.032 (-0.07)	-0.061 (-1.01)	0.28

(1) Log linear equation, all others are linear.
*Statistically significant at the 5% level or above.
**Statistically significant at the 10% level or above.
Value of the "t" statistic in parentheses.

Table 7.14. Market Share Equation: Corn, Quarterly Data

Import market	Constant	Lagged share	Price	R2
China (PRC)	-0.014 (-0.05)	0.795* (9.12)	-0.060 (-0.27)	0.65
EC	0.322 (2.97)	0.500 (3.67)	-0.007 (-0.07)	0.26
Egypt (1)	1.154 (1.48)	0.427* (4.19)	-12.045* (-3.39)	0.41
Japan	1.061 (6.06)	0.177 (1.32)	-0.373* (-3.28)	0.24
Mexico	1.659 (5.73)	0.263* (2.19)	-1.275* (-3.86)	0.34
South Korea (1)	-0.033 (-0.09)	0.657* (9.55)	-0.348 (-0.23)	0.67
Taiwan	0.214 (0.50)	0.664* (5.96)	0.055 (0.15)	0.46
USSR	0.232 (0.77)	0.698* (7.28)	0.010 (0.04)	0.61
USSR 1970-79			-0.009 (-0.03)	
USSR 1980-83			-0.095 (-0.32)	

(1) Log linear equation, others are linear.
*Statistically significant at the 5% level or above.
Value of the "t" statistic in parentheses.
All equations estimated with seasonal dummy variables.

Table 7.15. Market Share Equation: Wheat, Quarterly Data

Import market	Constant	Lagged share	Price	R2
China (PRC)(1)	-1.880	0.824*	-0.116	0.69
	(-2.05)	(10.23)	(-0.24)	
EC	0.142	0.048	-0.005	0.19
	(2.12)	(0.33)	(-0.07)	
Egypt (1)	1.007	0.887*	-4.973*	0.90
	(2.33)	(18.77)	(-1.50)	
Japan	0.853	0.179	-0.435**	0.16
	(3.04)	(1.31)	(-1.65)	
Mexico	1.989	0.535*	-1.768*	0.43
	(2.96)	(4.68)	(-2.50)	
South Korea	1.485	-0.025	-0.501*	0.13
	(5.50)	(-0.18)	(-2.21)	
Taiwan	1.284	0.423*	-0.674*	0.32
	(3.77)	(3.69)	(-2.73)	
USSR	1.141	0.614*	-0.934*	0.51
	(3.12)	(6.12)	(-2.68)	
USSR 1970-79			-0.883*	
			(-2.48)	
USSR 1980-83			-0.918*	
			(-2.69)	

(1) Log linear equation, others are linear.
*Statistically significant at the 5% level or above.
**Statistically significant at the 10% level or above.
Value of the "t" statistic in parentheses.
All equations estimated with seasonal dummy variables.

Table 7.16. Estimated Response of U.S. Market Share to Price

Market	Short-run market share effect (1)	Long-run market share effect (2)
Egypt corn	0.132	0.309
Japan corn	0.004	0.004
Mexico corn	0.010	0.040
Egypt wheat	0.049	0.055
Japa wheat	0.004	0.004
Mexico wheat	0.016	0.031
South Korea wheat	0.006	0.006
Taiwan wheat	0.009	0.021
USSR wheat	0.010	0.016

(1) Increase in share (proportion of total imports) during the current quarter from a 1% reduction in the ratio of U.S. to competitor prices.
(2) Long-run equilibrium increase in share from a 1% reduction in the price ratio, implied by the partial adjustment coefficient.

Table 7.17. Ranking of U.S. Markets based upon the Market Share Model

Markets with significant price response	Market share effect (1)	
1. Egypt corn	0.309	
2. Egypt wheat	0.055	
3. Mexico corn	0.040	
4. Mexico wheat	0.031	
5. Taiwan wheat	0.021	
6. USSR wheat	0.016	
7. South Korea wheat	0.006	
8. Japan wheat	0.004	
9. Japan corn	0.004	
Markets with lagged share effect but no price response		**Lagged share coefficient**
10. China wheat		0.824
11. China corn		0.795
12. USSR corn		0.698
13. Taiwan corn		0.664
14. South Korea corn		0.657
Markets with no price response or lagged share effect		
15. EC corn		
16. EC wheat		

(1) Long-run change in U.S. market share (proportion of total imports) associated with a 1% reduction in the ratio of U.S. to competitor prices.

NOTES

1. Leamer and Stern develop the directly-estimated demand equation for the case in which imports and domestic production are imperfect substitutes. Cronin has shown how the same assumption can be incorporated in the indirectly-derived excess demand equation. However, both of these approaches assume that imports from different sources are perfect substitutes for one another.

2. As Leamer and Stern point out, estimates of the total elasticity of demand for imports can be derived from the elasticity of substitution. Because of the small number of countries for which the model appears to be valid this was not done.

REFERENCES

Atkin, M. and D. Blandford. "Structural Change in Import Market Shares for Apples in the United Kingdom." European Review of Agricultural Economics 9(1982):313-326.

Cronin, M. R. "Export Demand Elasticities with Less Than Perfect Markets." Australian Journal of Agricultural Economics 23(1979):69-72.

Dent, W. T. "Application of Markov Analysis to International Wool Flows." Review of Economics and Statistics 44(1962):300-324.

Figueroa, E. E. "Implications of Changes in the U.S. Exchange Rate for Commodity Trade Patterns and Composition." Unpublished Ph.D. dissertation, University of California, Davis, 1986.

Goddard, E. W. "Export Demand Elasticities in the World Market for Beef." Mimeographed. University of Guelph, 1987.

Leamer, E. E. and R. M. Stern. Quantitative International Economics. Chicago: Aldine Publishing Co., 1970.

Lee, T. C. , G. G. Judge, and A. Zellner. Estimating the Parameters of the Markov Probability Model from Aggregate Time Series Data. Amsterdam: North-Holland, 1970.

Sirhan, G. and P. R. Johnson. "A Market Share Approach to the Foreign Demand for U.S. Cotton." American Journal of Agricultural Economics 53(1971):593-599.

Chapter 8

Ellen W. Goddard

Export Demand Elasticities in the World Market for Beef[1]

INTRODUCTION

The development and increasing use of international trade models for agricultural commodities has been in response to the realization that forecasting and policy analysis cannot be considered in a single country framework. In most scenarios for a country that exports its product the critical parameter which will affect the outcome of different policies is its export demand elasticity.

The determination of export demand elasticities facing a particular country has received a great deal of attention in the agricultural economics literature. The bulk of the analysis has rested on the assumption that goods produced in different countries are perfect substitutes. That this assumption may be false was recognized by Horner (1952) but Taplin (1971) was the first to attempt to quantify export demand elasticities in the presence of "product" (produced by an individual country) heterogeneity. To simplify his analysis he derived the export demand elasticity as a function of the aggregate elasticity of demand for the commodity (products produced in all regions) and cross-price elasticities (between products produced in different regions) assuming zero supply elasticities by competing exporters.

In this paper a framework for the determination of export demand elasticities, under the assumption that goods produced in different countries are not perfect substitutes, will be derived. The implications of the assumption for model structure and for estimation requirements will be described. The framework will be applied to results obtained in a model of international trade in beef. The assumption of perfect or imperfect substitutability by country of origin is usually a maintained rather than tested hypothesis. To show the implications of the assumption for the international beef market export demand elasticities will also be calculated for the market under an assumption of perfect substitutability.

MODELLING TRADE WITH IMPERFECT SUBSTITUTES

Goods exported from two different countries may not be perceived as perfect substitutes for each other or for the domestically produced good in an importing country for two main reasons. The first is the actual heterogeneity of the good in question. The second is "perceived or actual barriers to trade between countries" or alternately "a difference among suppliers rather than in product characteristics" (Johnson, Grennes and Thursby, 1979).

Barriers or perceived barriers to trade can exist in many forms, for example:

- Physical distances between exporters and importers generate:
  -direct per unit transport costs, and
  -indirect costs in the form of time lags.
- Reliability of suppliers, importer's wish to diversify source of imports.
- Economic and psychic distances between countries.
- Bilateral trade agreements.

Specification of an international trade model assuming product homogeneity by country of origin might have the following familiar structure shown in Table 8.1. The structure assumes no trade barriers or transportation costs and a common currency. This basic model solution is for a single world market clearing price (implicit in the assumption of product homogeneity). Regions either import or export the commodity at any single point in time.

Following Horner (1952) export demand elasticities facing each of Regions 3 and 4 can be determined as follows:

Exports from Region i = Demand in the Rest of the World -
Supply in the Rest of the World

$$\text{e.g., } E_3 = D_1 + D_2 + D_4 - S_1 - S_2 - S_4$$

$$\text{and } \quad E_4 = D_1 + D_2 + D_3 - S_1 - S_2 - S_3.$$

In general terms this can be expressed as:

$$E_i = \sum_{j=1}^{n} D_j - \sum_{j=1}^{n} S_j \qquad j = \text{number of other regions in the market.}$$

The export demand elasticity can be expressed as:

$$\frac{\partial E_i}{\partial P_i}\frac{P_i}{E_i} = \sum_{j=1}^{n}\frac{\partial D_j}{\partial P_i}\frac{P_i}{E_i} - \sum_{j=1}^{n}\frac{\partial S_j}{\partial P_i}\frac{P_i}{E_i}$$

or as (1)

$$\frac{\partial E_i}{\partial P_i}\frac{P_i}{E_i} = \sum_{j=1}^{n}\frac{\partial D_j}{\partial P_i}\frac{P_i}{D_j}\frac{D_j}{E_i} - \sum_{j=1}^{n}\frac{\partial S_j}{\partial P_i}\frac{P_i}{S_j}\frac{S_j}{E_i} \qquad j=1,\ldots,n \text{ for all } i.$$

The magnitude of the export demand elasticity (in absolute value terms) depends on the share of the exporter in supplies and demands of other regions in the market. Modifications to the above formula to account for barriers to trade, transportation costs, and domestic policy intervention can be found in Horner (1952) and Bredahl, Meyers and Collins (1979).

If we assume a four region model where goods produced in different regions are not perfectly substitutable, it is conceivable that every country might export to every other country. Thus, the model structure can be portrayed as in Table 8.2. Supply in each region is expressed as a function of the domestic price. Aggregate demand in each region is expressed as a function of a quantity weighted average price. The demands in each region for the goods produced in every region are expressed as a function of all four domestic prices (for example, D_{11} refers to demand in region 1 for good produced in region 1, D_{12} refers to demand in region 1 for good produced in region 2). Each regional model is closed with an identity equating supply to the sum of domestic demand and exports (determining regional price). The model expressed in this form has forty endogenous variables and equations (although the import and export identities in each region are not necessary, reducing the essential endogenous variables to 32). It is of interest to note that there is no single world market clearing price. However, the closer the elasticities of substitution are to infinity in each region the smaller the differences between prices in the various regions will be.

It should be pointed out that the model in Table 8.2 is based on a critical assumption; that each country exports a homogeneous commodity. In certain instances it is clear that the assumption that exporters export a homogeneous commodity is, in fact, no more defensible than the assumption that importers import a homogeneous commodity. However, for a variety of reasons, including most significantly, that a lot of the justification for imperfect substitutes on the demand side depends on country to country agreements/differences/risks in this research the assumption of product homogeneity on the supply side can be a maintained hypothesis.

Export demand elasticities can be derived in an international trade model of the above type as either partial or total elasticities. For example, if we describe exports from region i as the sum of demands in other regions for the good produced in region i or:

$$\begin{aligned} E_i &= D_{2i} + D_{3i} + D_{4i} \\ &= D_{2i}(P_1,P_2,P_3,P_4) + D_{3i}(P_1,P_2,P_3,P_4) + D_{4i}(P_1,P_2,P_3,P_4) \end{aligned}$$

the partial export demand elasticity (ε_i) facing region i is:

$$\varepsilon_i = \frac{\partial E_i}{\partial P_i}\frac{P_i}{E_i} = \left(\frac{\partial D_{2i}}{\partial P_i} + \frac{\partial D_{3i}}{\partial P_i} + \frac{\partial D_{4i}}{\partial P_i}\right)\frac{P_i}{E_i}, \quad \text{or}$$

$$\varepsilon_i = \sum_i k_i e_{ij}, \qquad i,j = 1,\ldots,n$$

$$\text{where } k_i = \frac{D_{ij}}{E_i}, \qquad D_{ij} = \text{quantity exported from i to a region j}$$

$$E_i = \text{total exports from i}$$

e_{ij} = partial demand elasticity for good from region i in region j

Following Buse (1958) the total export demand elasticity facing region i can be derived as:

$$TE_i = \varepsilon_i + \sum_{j=1}^{n} B_{ij} P_{ji}, \qquad n = \text{number of countries} \qquad (2)$$

where

ε_i = partial export demand elasticity derived above,

B_{ij} = cross price elasticity of demand for product produced in i with respect to the price of product produced in j

$$\text{(e.g. for Region i} = \left(\frac{\partial D_{2i}}{\partial P_j} + \frac{\partial D_{3i}}{\partial P_j} + \frac{\partial D_{4i}}{\partial P_j} \right) \frac{P_j}{E_i} \text{)}$$

P_{ji} = effect of a 1 percent change in the price of the product produced in i on the price of the product produced in j.

$$P_{ji} = \frac{S_{ji} - B_{ji}}{\varepsilon_j - s_j}$$

where S_{ji} = cross elasticity of supply of product produced in j with respect to price of product produced in i (zero in a trade context)

B_{ji} = cross elasticity of demand for product produced in j with respect to price of product produced in i

ε_j = partial export demand elasticity for product produced in j

S_j = supply elasticity for product produced in j.

With this formulation of total demand elasticity, if competing commodities are involved $|TE_i|$ will be less than $|\varepsilon_i|$ because B_{ij} and P_{ji} will both be positive. Alternately if goods are complementary the total demand elasticity may be more elastic than the partial demand elasticity. In a stable market where direct elasticities exceed cross elasticities $|\varepsilon_i| > |B_{ij}|$ and since $P_{ji} < 1$, TE_i will be negative.

It is important to note that, in comparing the two models discussed above, the critical difference is in the assumption of product homogeneity across all regions. Strictly speaking this assumption is not testable without perfect data. Ability to obtain consistent price and exact transportation cost data for every trade flow in a particular market might provide a basis for an a priori evaluation of perfect or imperfect substitution, i.e. if prices are the same at every point in time then the goods are identical. Perfect homogeneity would rule out any possibility of estimation of the second stage of the demand system portrayed in Table 8.2.

In a practical sense, for most agricultural commodities at a reasonable level of aggregation, most researchers would not deny that the assumption of product homogeneity is unrealistic. This stems from the biological nature of production for most agricultural commodities and the impact of exogenous factors such as weather and climate. Use of models with a maintained hypothesis of product homogeneity across countries for policy analysis purposes must be considered as an approximation to the real world where product heterogeneity is a fact.

The beef example utilized in the following empirical work is an attempt to illustrate what possible errors of direction or magnitude might be generated in a model with the maintained hypothesis of perfect substitution.

AN EMPIRICAL MODEL OF WORLD TRADE IN BEEF

World Beef Market

Of all the meats traded, beef is by far the most significant. The United States is the world's largest beef market with the highest level of production and imports. The USSR is also a large producer and sporadically a large importer. A fairly small percentage of world beef production enters world trade. Since 1965 the percentage has ranged from a low of 7 percent to a high of 11 percent. Apart from intra-EEC trade the world's major beef exporters are consistently the southern hemisphere countries of Australia, New Zealand and Argentina. In these countries a very high proportion of domestic production is exported (as high as 40-50 percent in Australia and New Zealand) making these countries particularly sensitive to changes in the international trading environment. It is interesting to note that most countries involved in international beef trade both export and import the commodity.

The international beef market is affected by a variety of barriers to trade including health regulations, import quotas, tariffs, variable levies and in some cases export control programs. The major barrier to trade for some market participants is the fact that some of the major importing nations refuse to allow imports from regions with foot and mouth disease. This largely segregates the world market into two sub-markets, the Pacific Basin with trade flowing from Oceania to the United States, Japan and Canada and the Atlantic region with trading flowing from South America to Western Europe. In the years when the USSR has been an importer of beef it has taken beef from both Australia and Argentina. The health regulations do not completely divide the beef market into two, in some of the importing regions of the world such as Southeast Asia and the Middle East, there is competition between the 'free' exports from Oceania and the exports from South America. Recently, however, the EEC has become a more significant exporter of beef and is replacing South America as a supplier of beef to some European and Middle Eastern countries. The EEC is also exporting to some traditional Oceanic markets such as Canada.

Most of the major import markets such as the United States, Canada, Japan and Korea have import quotas on beef. In North America the import restrictions are generally achieved through voluntary restraint agreements with exporting countries, generating a situation where import quotas themselves do not have to be activated. This voluntary restraint mechanism allows the economic rent accruing to the import quota to be captured in exporting countries, making these particular export markets even more desirable. The exporting countries themselves are faced with the necessity of regulating the quantities of beef flowing to these import markets. Australia has handled this problem by granting export rights to the United States to particular exporters on the basis of their historical percentage of total exports to uncontrolled markets, or more recently, to all markets. This policy has encouraged/forced exporters in Australia to be aggressive in international beef markets and driven up domestic beef prices. The Japanese and Korean import quotas are not handled with a voluntary restraint mechanism and Australia, New Zealand and the United States are all competitors for a share of these markets.

Production systems for beef vary between regions in the world. Northern hemisphere beef is produced through feedlots with cattle being fattened on grain and protein concentrates. In the southern hemisphere cattle are generally grass fed in a rangeland type of production. The quality of beef produced in each system can be quite variable. There can also be a strong link between beef and dairy production. In most countries beef is produced from dairy herds in the form of cull cows and calves. The relative importance of this sort of beef can be significantly affected by dairy policy, a phenomenon currently being exhibited in the United States and the EEC. In other countries, i.e. Japan and Korea, dairy calves are fattened for slaughter providing a significant percentage of total beef production. Even in these countries, however, there is an observable quality difference between mature beef from dairy cattle and that from beef cattle.

If one is interested in modelling the international beef market for policy analysis or forecasting purposes there are some key characteristics to be considered:

- most countries export and import beef;
- the quality of beef produced in different countries from different types of cattle and from different production systems can be quite variable;
- the market is partially segmented by health regulations but parts of the market cannot be considered in isolation of each other;

- to major export countries the international market is critical to the livelihood of the domestic industry.

Empirical World Beef Model

The above characteristics of the international beef market necessitate the specification of an international beef model with product heterogeneity as in Table 8.2. However, the trade model portrayed in Table 8.2 is devoid of information relating to functional form and the exact specification of different equations within a single region's model. In empirically estimating such a model it is essential to specify a consistent framework that can be applied where relevant across all regions involved in the world market.

The bulk of the previous econometric analysis of beef markets has been conducted in a one or two country framework. However, these previous studies provide a wealth of information with respect to modelling of livestock markets. Traditional livestock market models have concentrated on capturing the dynamics of the production process. The supply component of a livestock market model is usually expressed as a series of simultaneous equations describing the actions of producers at various stages in the production process. Traditionally, demand has been expressed as a per caput consumption equation combined with price linkages across different levels of the marketing chain. A general model for each region in the world beef market is presented below in its two major components, first supply and then demand.

Supply

The major distinguishing characteristic of supply of a livestock commodity when compared to that of a crop is that the product of the process, i.e. cattle, are at one and the same time both capital and consumption goods. Other critical characteristics include the length of the beef production process (at least three years from the decision to breed a cow until a steer is slaughtered) and the fact that climate can play a critical role in determining the nature of the production process (range-feeding or feedlot-feeding). Jarvis (1974) has developed the most elegant and certainly, the most definitive theoretical framework for the supply side of an econometric model of a cattle sector. His framework captures the dual nature of cattle as well as the interrelationship between the dairy and beef sectors. With some modifications his framework has been used in most previous econometric studies (see, for example, Freebairn and Rausser; Ospina and Shumway). The framework he has developed is utilized in this study.

Differences in the supply models specified in this study occur due to data limitations as well as to more critical structural considerations. For regions where beef is generally raised in feedlots (U.S., Canada) feed prices are critical explanatory variables. In regions where cattle are range-fed, feed prices are not included as explanatory variables. Optimally, cattle slaughter (beef supply) should be disaggregated by sex (in a range-feeding system) and by age (in a feedlot-feeding system). This disaggregation is only possible where complete data is available. Thus, there is some variation in the structure of the supply component across regions in the model. The simplest possible supply side specification is shown in Table 8.3. This specification is used for all aggregate regions (composed of more than one country) and for all countries where disaggregated data could not be obtained. A priori the sign on the price of beef in the beef production equation is expected to be negative while that for expected price of beef in the cattle inventory equation is expected to be positive. For some critical regions in the world beef market, e.g. Australia, U.S., Canada and Japan, beef production is disaggregated into a series of equations explaining slaughter of animals by sex and/or age and carcass weights. Whatever the regional supply component of the model each region is assumed to produce a single product, beef.

Demand

The demand side of most agricultural commodity models takes the form of per caput consumption equations relating consumption to retail prices and income as well as price linkages between farm or wholesale and retail prices. This traditional approach is sufficient to satisfy the requirements of a trade model assuming product homogeneity across regions (Table 8.1). However, the requirements of a model allowing for product heterogeneity by country of origin are somewhat different. Essentially, using traditional consumer theory, demand for an individual product can be expressed as a function of all prices and income. Allowing for product heterogeneity by country of origin increases the number of prices that might be included in a single demand equation by the number of regions in the analysis.

In the seminal article by Armington (1969) the approach to "simplifying the product demand functions to the point where they are relevant for the practical purposes of estimation and forecasting" was to impose a series of restrictive assumptions. These assumptions were that of

- a two stage budgeting process and
- a linear homogeneous utility function of a CES form.

The CES functional form implies that elasticities of substitution between products of the same kind in a market are constant and are the same between any pair in the market as between any other pair in the same market. The linear homogeneous utility function at the second stage of the process is a necessary condition of consistent two stage budgeting (Green, 1971). The two stage demand system approach is, to date, the only practical solution to the modelling of demand in an international framework where products are differentiated by country of origin.

The two stage demand model can be applied to a regional beef model as shown in Table 8.3. In the first stage aggregate consumption of the commodity, beef, is determined. In the second stage, consumption of beef from different sources is determined. A necessary requirement for a two-stage demand system is that the marginal rates of substitution between goods at the second stage (beef from different sources) are independent of the quantities of other meats, of other foods and of all other goods consumed.

The first stage is made up of a single equation explaining per caput consumption (demand) of beef as a function of prices of beef and substitute meats and per caput income. Ideally this stage would be composed of a system of equations explaining the allocation of total income over a complete range of aggregate commodity groupings. This is not possible within the confines of this study. The price of beef used in the aggregate consumption equation is the quantity weighted average of the prices of beef from different sources. Total expenditure on beef is derived as the product of the weighted average price and aggregate consumption.

The second stage is made up of a system of equations explaining the demand for beef from different sources as functions of prices of beef from all sources and total expenditure on beef. The approach used is to specify an indirect utility function with a Generalized Box-

Cox functional form and appropriate derived system of demand equations. This approach differs from that used by Armington in which a CES functional form was used for the second stage of the demand system. As pointed out previously, the use of a CES functional form entails strict restrictions on the elasticities and relationships between the goods in the demand system. To reduce the number of restrictions on the system it was decided to use one of the class of, recently popular, flexible functional forms proposed for both utility and/or production functions (flexible in that they do not, a priori, constrain price, income or substitution elasticities). The Generalized Box-Cox functional form is appealing in that it contains many other flexible functional forms as special cases which can be tested for statistically. This characteristic is particularly appropriate when theory cannot provide any specific information about the underlying functional relationships.

The share equations (e.g., value of beef from each source as a proportion of total expenditure on beef) derived by Roy's identity from the indirect utility function are as portrayed in Table 8.3. The demand component of a regional beef model is thus composed of an aggregate per caput beef consumption equation, a total expenditure identity and a system of expenditure share equations (for beef from all sources including the domestic source). Trade barriers of all kinds including transportation costs, tariffs, voluntary export restraints can be incorporated into the specification (see Goddard 1984). Import quotas should conceptually be handled by incorporating an extra restriction into the consumer choice problem (and eventually an extra variable in the expenditure share system and per caput consumption equation). They are handled in this research by separating the decision to consume beef (expenditure share system is only estimated over beef from import sources) into two decisions to consume beef from domestic and imported sources. If a region does not import beef from any source (e.g. the major exporting countries Australia, New Zealand, Argentina) then the demand component of the model reduces to the aggregate per caput beef consumption equation.

Each regional beef model is closed by an identity equating the sum of demands in every region including the domestic one with production.

Empirical Estimation

Regional beef models are estimated for the following countries and regions:

- Canada, U.S., Australia, New Zealand, Japan, South Korea,
- Argentina, Uruguay, Brazil
- Mexico - Central America - Caribbean
- Rest of South America
- European Community
- Western Europe
- Africa
- North Africa - Middle East

The major exclusions from the above list are the U.S.S.R., Eastern Europe and parts of Asia. It was found too difficult to obtain data to model these regions endogenously. Eastern Europe does appear in the model however as an exogenous supplier of beef to the other European regions. Beef supply, demand and price data used in the analysis was taken from published domestic sources (i.e., Australia Meat and Livestock Corporation, U.S.D.A., Agriculture Canada, Japan Ministry of Agriculture) and international sources (O.E.C.D. Meat Balances, FAO Production Statistics). All volume and value data on trade flows was taken from the United Nations trade statistics. The sample period for the analysis was 1960-1980.

One problem that arose when estimating the model centered around the determination of trading partners for the second stage of the demand systems in importing countries/regions. The only trading partners included in the estimation were those partners that were consistent exporters over the sample, 1960-1980. While this may bias the results obtained Richardson (1973, p. 385) notes that it is not clear whether the exclusion of these trading partners biases the estimates of the price or substitution elasticities.

From the general model structure specified in Table 8.3 it is clear that supply, demand and prices are determined simultaneously in all regions in the model. It is well known, that "estimation of a system of simultaneous equations by ordinary least squares produces estimates that are biased and inconsistent because of the inclusion of the endogenous variables among the set of explanatory variables" (Intriligator, 1978, p. 374). There are however cases in which OLS may be a more appropriate estimator than other simultaneous equation estimators. These include the following:

- if the model is recursive or approximately recursive
- if the matrix of coefficients of the endogenous variables is sparse
- if there is substantial multicollinearity among exogenous variables (not all of which might be included in one equation) (Intriligator, 1978; Johnston, 1972).

For these reasons the major estimator used in the beef model is OLS. The second stage of the demand systems, in relevant countries or regions is estimated by seemingly unrelated regression techniques to account for the dependence among error terms and to allow cross-equation coefficient constraints. All equations are assumed to have additive error terms. Statistical hypothesis testing is conducted by means of t and F tests in the single equations and with likelihood ratio tests in the second stage of the demand systems. In some cases restrictions of consumer theory, i.e., linear homogeneity of indirect utility functions, were statistically rejected. They were not imposed on the demand systems and further results were established without restrictions. Symmetry, however, was imposed in all cases, even in the few cases where it was statistically rejected. Results from individual equation estimates can be found in Goddard (a,b,c,d).

The complete model (300 endogenous variables and equations) was simulated for validation purposes over the last ten years of the estimation period (1970-1980) using a Newton algorithm for the solution of nonlinear simultaneous equations. All computer work was done using TSP version 4.

Model Validation Results

As mentioned previously much of the modelling work done on the beef sector has been in a one or two country framework. For example, the U.S. beef market has had extensive modelling efforts directed at it. To a lesser extent the Canadian and Australian beef sectors have received attention. The merits of the current beef model over those of the previous models are difficult to ascertain on a straight comparison basis. The models concentrated on capturing different aspects of the market. The basic aim of this research was to establish whether beef from different countries of origin could be modelled in a perfect substitution framework or not. While this hypothesis is essentially non-testable a comparison of the fit of trade variables modelled under the assumption of perfect substitution when compared to the fit of trade variables modelled under the alternate assumption might provide some evidence to support or reject a particular hypothesis. Given the partial (in terms of country coverage) analysis of all previous models such a comparison is not possible.

The complete model was validated using historical exogenous data as well as under a number of different shocks. As well, individual regional models were validated. These results are reported in a number of related publications (Goddard, a,b,c,d). On a variety of criteria the model was deemed acceptable for policy analysis purposes.

THE MEASUREMENT OF EXPORT DEMAND ELASTICITIES

It is possible to simulate the estimated international beef model under a variety of exogenous supply shocks to individual beef exporters to ascertain implicit export demand elasticities. As well, it is possible to unhook the models of demand for beef by source and resimulate the entire model to establish the range of export demand elasticities which might have been generated under an assumption of perfect substitutability by country of origin. This exercise is not conducted with the model for a number of reasons. The first and most important is that the size and complexity of the model make the comparison simulations difficult to engineer. Various equations in the existing model i.e., the per caput consumption equations would have to be reestimated to establish relevant parameters. As well, the results would not be directly comparable since the perfect substitution model would only generate net trade data for each region, not the trade flow data that is the crux of the imperfect substitution model. The second major reason is that the significant differences in the specification of the supply components of the model across regions make it difficult to establish "shocks" for each one of the regional models that would be consistent.

The major point of this exercise was to establish whether the assumption of imperfect substitution in the world beef market leads to significantly different export demand elasticities than would have been generated otherwise. To determine this it is possible to use the formulas established previously (1, 2) and specific estimated supply and demand elasticities to calculate export demand elasticities under each assumption.

The elasticities relevant for the derivation of export demand elasticities are reported in Tables 8.4 through 8.7. Standard errors are not reported for all elasticities as they are complex involving the 50-100 parameters resulting from the second stage demand systems. Aggregate demand elasticities (see Table 8.4) are inelastic for all regions. Japan and Australia have the largest price elasticities (in absolute value terms).

The reported supply elasticities in Table 8.5 show a wide degree of variation. Canada, for example, has a very large supply elasticity while Australia has a very small one. In both cases it may be possible that internal policies (e.g., stabilization programs, research) may have affected the supply elasticities. Much testing of different specifications (functional forms, linear and logarithmic and different lags on expected price series) was undertaken and the various supply elasticities were robust across different specifications.

Own and cross price elasticities for beef by country of origin are reported in Table 8.6. These elasticities are derived from the second stage of the demand systems specified earlier. The elasticities are not constant and are reported for the sample midpoint. Many cross price elasticities are zero showing the relative sparseness of the trade flow matrix over some or all of the estimation period. The table can be interpreted as follows. The own price elasticity of demand for Australian beef in Japan is -1.02. The cross price elasticity of demand for Australian beef in Japan with respect to beef from New Zealand is -.070 (implying gross complementarity). The cross-price elasticity of demand for New Zealand beef in Japan with respect to the price of Australian beef is -.055 (implying gross complementarity). The fact that Japan has only consistently imported beef from Australia, New Zealand and the U.S. over the sample period is reflected in the fact that all other cross price elasticities (e.g., for Argentinian beef) are zero.

In equations (1) and (2) reported previously expressions for the determination of export demand elasticities under differing assumptions of elasticities of substitution were presented. Both of these equations can be applied to the international beef market supply and demand elasticities reported. Using equation (1) the data required are the aggregate supply and demand elasticities and quantities exported, consumed and produced in each region in the world market. Export demand elasticities calculated under this assumption are presented for 1972 and 1980 in columns one and two in Table 8.7. As expected, a priori, export demand elasticities are smaller for larger exporting nations, e.g., Australia, than for small exporters, e.g., U.S. The elasticities are smaller, on average, for the South American countries since they are not allowed to export to North America or Japan due to foot and mouth disease regulations. These reported elasticities suggest that countries involved in world beef trade face very elastic export demand curves.

Combining the relevant second stage demand elasticities, the supply elasticities and equation (2) partial and total export demand elasticities assuming imperfect substitutability can be calculated and are reported in Table 8.7. In most cases (Australia, Canada, EC-10, Rest of Western Europe, Argentina, Brazil, Uruguay, Rest of South America) the total export demand elasticities are slightly smaller than the partial elasticities. For New Zealand beef and beef from the U.S. there are very small increases (in absolute value terms) as you move from partial to total elasticity, suggesting complementarity with beef from other sources. These phenomenon may arise from institutional characteristics of the international beef market, e.g., the distribution of imports in the regulated Japanese market and voluntary export restraints in the U.S. and Canada. Simulation of the regional Japanese model, for example, under a 10 percent increase in total imports resulted in an increase in imports from the U.S. of 13 percent, from Australia of 10 percent and from New Zealand of 7 percent. The complementarity of beef from New Zealand and the U.S. with that from other sources implies a slight positive impact on exports from the two countries as exports from other sources increase.

Most export demand elasticities are approximately one with the exception of Western Europe (≈ 5), the U.S. and the Rest of South America (≈ 2). The Rest of Western Europe's major export market is the EC. The results suggest that the EC imports from the Rest of Western Europe are very responsive to their price changes. This results from the institutional regulations surrounding beef trade to the EC. Other exporters to the EC face more inelastic demands due to arrangements like the LOME agreement.

The international beef market is affected by a number of institutional arrangements. Health regulations in major importing countries (U.S., Japan) ban trade from South America segregating but not isolating the world market into Pacific and Atlantic Basins. Voluntary export restraints with distribution of export rights established on a historical basis limit access to major markets like

Canada and the U.S. Japan allocates import rights on a bilateral basis with the result that U.S. exports to Japan have been growing over time. All of these arrangements have an impact on the export demand elasticity facing a particular country. They have all been incorporated into the specification and estimation of the model and, in turn, into the calculated export demand elasticities. What a researcher would like to establish at this juncture, for a specific exporter, is the key variable affecting his export demand elasticity i.e., is a change in U.S. beef import regulations likely to have an impact on the export demand elasticity facing New Zealand. The myriad of international arrangements as well as domestic distortions (e.g., EC-10 Common Agricultural Policy) make it impossible to establish the critical factor determining a particular export demand elasticity.

If the results in Table 8.4 are compared to those in Table 8.7 it is noticeable that in every case the export demand elasticity is greater than the domestic aggregate demand elasticity. The fact that beef exporters face different, but inelastic, demands on both domestic and export markets opens up the opportunity for exporting countries to pursue marketing strategies (for example, price discrimination) that may increase their profits associated with selling/trading beef.

It is interesting to note that even very small exporters, such as Canada, face export demand elasticities which differ only slightly from those of their much larger counterparts. This is a result that is not necessarily obvious from the traditional "homogeneous good" definition of export demand elasticities.

CONCLUSION

An incorrect assumption about whether products from different countries are perfect substitutes in international markets biases estimates of export demand elasticities significantly. Market strategy for both domestic and international markets is affected by elasticities of supply and demand. Optimal marketing strategies established for particular countries may be inadequate if the elasticities are biased.

Export demand elasticities established for the international beef market under an assumption of imperfect substitutability, are all close to 1. They differ from domestic demand elasticities which in all cases are less than 1 and in most cases less than .5. These elasticities suggest the scope for beef exporters to pursue aggressive marketing strategies rather than operate as if they were price takers on world markets. The implications, however, of pursuing specific marketing strategies must be examined, a priori, within the context of the structural international beef model. This would ensure that the constraints in the market (quotas, voluntary export restraints, health regulations, etc.) would be captured explicitly.

If some of the structural estimates of the international beef model are used to establish export demand elasticities under the assumption of perfect substitutability, these elasticities are very large for all exporters. Similar results would likely be obtained with any other international beef model that rests on an assumption of perfect substitutability.

While the estimated results reported in this study cannot provide any direct evidence as to the cause of the magnitude of the export demand elasticities they provide a critical input to the modelling decisions of a researcher. Relevant data must be analyzed thoroughly, a priori, to ascertain whether the assumption of infinite elasticities of substitution is realistic or not. While the move to a structural model that allows for imperfect substitutability increases the data requirements and complexity significantly it may be necessary if meaningful policy analysis and forecasting tools are to be generated.

Table 8.1. Trade Model (Without Product Differentiation)

Region 1	Region 2	Region 3	Region 4
Supply = $f(P_1)$ Demand = $f(P_1)$	Supply = $f(P_2)$ Demand = $f(P_2)$	Supply = $f(P_3)$ Demand = $f(P_3)$	Supply = $f(P_4)$ Demand = $f(P_4)$
Imports1=Demand-Supply	Imports2=Demand-Supply	Exports3=Supply-Demand	Exports4=Supply-Demand
	Imports1 + Imports2 = Exports3 + Exports4		
	P_1=P_2=P_3=P_4		
Endogenous Variables			
Supplies(4)	Demands(4)	Imports/Exports(4)	Prices(1)

Table 8.2. Trade Model (With Product Differentiation)

Region 1	Region 2	Region 3	Region 4
$Supply = f(P_1)$	$Supply = f(P_2)$	$Supply = f(P_3)$	$Supply = f(P_4)$
$Demand = f(WP_1)$	$Demand = f(WP_2)$	$Demand = f(WP_3)$	$Demand = f(WP_4)$
$D_{11} = f(P_1,P_2,P_3,P_4)$	$D_{21} = f(P_1,P_2,P_3,P_4)$	$D_{31} = f(P_1,P_2,P_3,P_4)$	$D_{41} = f(P_1,P_2,P_3,P_4)$
$D_{12} = f(P_1,P_2,P_3,P_4)$	$D_{22} = f(P_1,P_2,P_3,P_4)$	$D_{32} = f(P_1,P_2,P_3,P_4)$	$D_{42} = f(P_1,P_2,P_3,P_4)$
$D_{13} = f(P_1,P_2,P_3,P_4)$	$D_{23} = f(P_1,P_2,P_3,P_4)$	$D_{33} = f(P_1,P_2,P_3,P_4)$	$D_{43} = f(P_1,P_2,P_3,P_4)$
$D_{14} = Demand - D_{11} - D_{12} - D_{13}$	$D_{24} = Demand - D_{21} - D_{22} - D_{23}$	$D_{34} = Demand - D_{31} - D_{32} - D_{33}$	$D_{44} = Demand - D_{41} - D_{42} - D_{43}$
$WP_1 = \frac{(P_1D_{11}+P_2D_{12}+P_3D_{13}+P_4D_{14})}{Demand}$	$WP_2 = \frac{(P_1D_{21}+P_2D_{22}+P_3D_{23}+P_4D_{24})}{Demand}$	$WP_3 = \frac{(P_1D_{31}+P_2D_{32}+P_3D_{33}+P_4D_{34})}{Demand}$	$WP_4 = \frac{(P_1D_{41}+P_2D_{42}+P_3D_{43}+P_4D_{44})}{Demand}$
$Imports = D_{12} + D_{13} + D_{14}$	$Imports = D_{21} + D_{23} + D_{24}$	$Imports = D_{31} + D_{32} + D_{34}$	$Imports = D_{41} + D_{42} + D_{43}$
$Exports = D_{21} + D_{31} + D_{41}$	$Exports = D_{12} + D_{32} + D_{42}$	$Exports = D_{13} + D_{23} + D_{43}$	$Exports = D_{14} + D_{24} + D_{34}$
$Supply = D_{11} + Exports$	$Supply = D_{22} + Exports$	$Supply = D_{33} + Exports$	$Supply = D_{44} + Exports$

Endogenous Variables
Supplies (4), Demand (4), Regional Demands (16), Regional Prices (4), Weighted Prices (4), Imports/Exports (8)

Table 8.3. Regional Model

Demand Model

(1) $Demand_{Beef} = A + B(WP_{Beef}) + C(PS) + D(Y) + u$

$Demand_{Beef}$ = per capita consumption of beef

PS = price of a substitute good, e.g., pork in some regions

Y = disposable income

WP_{Beef} = quantity weighted average price of beef

(2) Total $Expenditure_{Beef} = Demand_{Beef} \times WP_{Beef}$

(3) Expenditure Shares on Beef from Different Sources =

$$X_i V_i = \left(\frac{\alpha_i V_i^\lambda + \sum_i \beta_{ij} V_i^\lambda \frac{(V_j^\lambda - 1)}{\lambda}}{\sum_i \alpha_i V_i^\lambda + \sum_i \sum_j \beta_{ij} V_i^\lambda \frac{(V_j^\lambda - 1)}{\lambda}} \right) + u$$

X_i = quantity of beef from i

P_i = price of beef from i

V_i = P_i / Total $Expenditure_{Beef}$

Supply Model

(1) Beef Production = E + F(Price Beef) + G(Cattle $Inventory_{t-1}$) + u

(2) Cattle Inventory = H + J(Expected Price Beef) + K(Cattle $Inventory_{t-1}$) + u

Model Closure

(1) Sum of Demands for Beef produced in region i by all j regions = Beef Production in i

---

$A,B,C,\alpha,\beta,\lambda,E,F,G,H,J,K$ = estimated parameters

u = regression errors

Table 8.4. Beef Price Demand Elasticities

	Aggregate Beef Price Demand Elasticity[1]	Own Price Demand Elasticity[2]
Canada	-.44	-.95
U.S.	-.45	-.90
Australia	-.63	
New Zealand	-.34	
EC-10	-.21	
Rest of Western Europe	-.20	
Africa	-.18	
North Africa - Middle East	-.40	
Mexico - Central America - Caribbean	-.34	
Argentina	-.46	
Uruguay	-.15	
Brazil	-.14	
Rest of South America	-.11	
Japan	-.85	-.90

[1]Calculated at sample means from per capita consumption equations.
[2]Calculated at sample means from expenditure share equations.

Table 8.5. Supply Elasticities for all Regional Models (defined as long run response in cattle inventories to price of beef)

	Beef Price Supply Elasticity[1]
Japan	.45
Canada	1.83
U.S.	.40
Australia	.12
New Zealand	.42
EC-10	1.00
Rest of Western Europe	.11
Argentina	.32
Uruguay	.40
Brazil	1.40
Rest of South America	.71
Mexico - Central America - Caribbean	.12
Africa	.41
North Africa - Middle East	.15

[1]Calculated at sample means.

Table 8.6. Own and Cross Price Elasticities of the Import Demand for Beef by Country of Origin

Exporters / Importers	Australia	New Zealand	U.S.	Canada	EC-10	M.C.A.	Argentina	Brazil	East Europe	Uruguay	Africa	Rest of W. Europe	Rest of S. America
Australian Beef													
Japan	-1.020	-0.070	0.074	0.000	0.000	0.000	0.000	0.000	0.000	0.000	0.000	0.000	0.000
Korea	-0.918	-0.080	-0.503	0.000	0.000	0.000	0.000	0.000	0.000	0.000	0.000	0.000	0.000
Canada	-1.660	0.910	-0.180	-0.700	0.000	0.000	0.000	0.000	0.000	0.000	0.000	0.000	0.000
U.S.	-0.922	0.100	-1.560	0.050	0.330	-0.220	0.000	0.000	0.000	0.000	0.000	0.000	0.000
Rest of W. Europe	-1.540	-1.260	0.003	0.000	-1.260	0.000	-0.002	0.940	1.460	-0.860	1.530	[0]	0.000
EC-10	-1.860	-0.558	----	0.000	[0]	0.000	0.191	0.395	1.360	0.000	-2.060	1.000	0.554
Rest of World	-1.000	0.000	0.000	0.000	0.000	0.000	0.000	0.000	0.000	0.000	0.000	0.000	0.000
New Zealand Beef													
Japan	-0.055	-0.635	0.435	0.000	0.000	0.000	0.000	0.000	0.000	0.000	0.000	0.000	0.000
Korea	0.206	-1.290	0.302	0.000	0.000	0.000	0.000	0.000	0.000	0.000	0.000	0.000	0.000
Canada	1.020	-1.410	0.720	-2.250	0.000	0.000	0.000	0.000	0.000	0.000	0.000	0.000	0.000
U.S.	0.230	-0.940	-1.820	-0.046	0.190	0.060	0.000	0.000	0.000	0.000	0.000	0.000	0.000
Rest of W. Europe	-4.880	-1.560	1.230	0.000	0.200	0.000	1.670	1.410	-1.680	1.590	0.480	[0]	0.000
EC-10	-2.460	-0.720	0.000	0.000	[0]	0.000	0.199	-0.039	0.000	5.620	-2.910	1.080	-0.664
Rest of World	0.000	-1.000	0.000	0.000	0.000	0.000	0.000	0.000	0.000	0.000	0.000	0.000	0.000
U.S. Beef													
Japan	0.207	0.214	-1.810	0.000	0.000	0.000	0.000	0.000	0.000	0.000	0.000	0.000	0.000
Korea	-1.020	1.350	-0.707	0.000	0.000	0.000	0.000	0.000	0.000	0.000	0.000	0.000	0.000
Canada	-0.250	1.000	-2.030	-0.700	0.000	0.000	0.000	0.000	0.000	0.000	0.000	0.000	0.000
Rest of W. Europe	0.330	2.240	-1.510	0.000	-2.680	-1.370	-0.410	-0.240	1.990	2.130	0.000	0.000	0.000
Rest of World	0.000	0.000	-1.000	0.000	0.000	0.000	0.000	0.000	0.000	0.000	0.000	0.000	0.000
Canadian Beef													
U.S.	0.500	-0.180	-1.630	-1.230	-0.610	1.150	0.000	0.000	0.000	0.000	0.000	0.000	0.000
Rest of World	0.000	0.000	0.000	-1.000	0.000	0.000	0.000	0.000	0.000	0.000	0.000	0.000	0.000

Rest of World elasticities have been assumed.

Table 8.6. Continued.

Exporters / Importers	Australia	New Zealand	U.S.	Canada	EC-10	M.C.A.	Argentina	Brazil	East Europe	Uruguay	Africa	Rest of W. Europe	Rest of S. America
Argentinian Beef													
Rest of S. America	0.000	0.000	0.000	0.000	0.000	0.000	-0.966	0.000	0.000	-0.023	0.000	0.000	0.000
Rest of W. Europe	-0.070	0.220	-0.130	0.000	0.590	0.000	-3.360	0.570	-0.200	1.010	-0.460	0.000	0.000
EC-10	0.032	0.016	0.000	0.000	0.000	0.000	-0.915	-0.011	-0.084	0.000	-0.029	-0.530	-0.017
Rest of World	0.000	0.000	0.000	0.000	0.000	0.000	-1.000	0.000	0.000	0.000	0.000	0.000	0.000
Brazilian Beef													
Rest of W. Europe	0.310	0.130	-0.030	0.000	0.300	0.000	0.490	-1.590	-0.003	0.020	-0.370	[0]	0.000
EC-10	0.323	0.038	0.000	0.000	[0]	0.000	-1.300	-0.882	-1.290	0.000	0.902	-0.434	0.254
Rest of World	0.000	0.000	0.000	0.000	0.000	0.000	0.000	-1.000	0.000	0.000	0.000	0.000	0.000
Uruguayan Beef													
Rest of S. America	0.000	0.000	0.000	0.000	0.000	0.000	-4.360	0.000	0.000	-7.070	0.000	0.000	0.000
Rest of W. Europe	-0.410	-0.190	-0.040	0.000	-0.470	0.000	0.850	-0.180	-0.490	-1.150	0.420	0.000	0.000
Rest of World	0.000	0.000	0.000	0.000	0.000	0.000	0.000	0.000	0.000	-1.000	0.000	0.000	0.000
Rest of S. American Beef													
EC-10	0.826	-0.238	0.000	0.000	0.000	0.000	-0.047	0.459	-0.479	0.000	0.000	-0.337	2.120
Rest of World	0.000	0.000	0.000	0.000	0.000	0.000	0.000	0.000	0.000	0.000	0.000	0.000	-1.000
EC-10 Beef													
U.S.	3.500	0.900	-0.220	-0.700	-1.960	-1.950	0.000	0.000	0.000	0.000	0.000	0.000	0.000
Rest of W. Europe	-0.460	0.030	-0.160	0.000	-1.040	0.000	0.560	0.340	0.440	-0.270	-0.270	0.000	0.000
Rest of World	0.000	0.000	0.000	0.000	-1.000	0.000	0.000	0.000	0.000	0.000	0.000	0.000	0.000
Rest of W. European Beef													
EC-10	4.070	0.934	0.000	0.000	0.000	0.000	-0.508	-1.790	3.490	0.000	-1.270	-5.330	-8.480
Rest of World	0.000	0.000	0.000	0.000	0.000	0.000	0.000	0.000	0.000	0.000	0.000	-1.000	0.000

Relationships in boxes have been constrained to zero due to the presence of quantitative trade barriers or regional aggregation.
Other zeros in the matrix arise because there was no consistent trade on that particular trade flow over the sample period.

Table 8.7. Partial and Total Export Demand Elasticities Facing Markets in International Beef Trade*

	Elasticities generated under the assumption of perfect substitutability		Elasticities generated under the assumption of imperfect substitutability			
	Total		Partial		Total	
	1972	1980	1972	1980	1972	1980
Australia	-36.74	-36.52	-1.125	-1.000	-0.981	-1.060
New Zealand	-102.65	-96.64	-0.986	-0.990	-1.308	-1.178
U.S.	-587.29	-396.08	-1.655	-1.712	-1.899	-2.099
Canada	-539.29	-400.52	-1.139	-1.220	-0.745	-1.073
EC-10	-374.089	-333.41	-1.408	-1.130	-1.096	-1.049
W. Europe	-1022.36	-728.62	-5.305	-5.313	-4.628	-5.219
Argentina	-26.60	-97.59	-1.044	-1.258	-1.016	-1.231
Brazil	-59.91	-2351.74	-1.115	-1.222	-0.979	-1.142
Uruguay	-213.21	-1130.48	-1.068	-0.904	-1.023	-0.887
Rest of S. America	-19.80	-21.50	-1.748	-2.021	-1.699	-2.001

*Calculated from demand and supply elasticities reported earlier.

NOTES

1. The author would like to acknowledge financial support from the International Trade Research Consortium and the Ontario Ministry of Agriculture and Food, and review comments from Karl Meilke, Colin Carter et al.

REFERENCES

Allen, R. G. D. (1938). Mathematical Analysis for Economists, MacMillan Press Ltd., London.

Armington, P. S. "A Theory of Demand for Products Distinguished by Place of Production," I.M.F. Staff Papers, (March and July 1969):159-177.

Bredahl, M.E., W. H. Meyers, and K. J. Collins. "The Elasticity of Foreign Demand for U.S. Agricultural Products: The Importance of the Price Transmission Elasticity," American Journal of Agricultural Economics 61(No. 1, 1979):58-61.

Buse, R. C. "Total Elasticities - A Predictive Device," Journal of Farm Economics 40(No. 4, November 1958):881-891.

Freebairn, J. W. and G. C. Rausser. "Effects of Changes in the Level of U.S. Beef Imports," American Journal of Agricultural Economics 57(1975):676-688.

Goddard, E. W. (1984). "Analysis of International Beef Markets," unpublished Ph.D. thesis, La Trobe University, Melbourne, Australia.

____________. (1983a). "Models of the Beef Market in Japan and Korea," Occasional Paper No. 3, School of Agriculture, La Trobe University.

____________. (1983b). "Models of the Beef Market in the U.S. and Canada," Occasional Paper No. 4, School of Agriculture, La Trobe University.

____________. (1983c). "Models of the Beef Market in Australia and New Zealand," Occasional Paper No. 5, School of Agriculture, La Trobe University.

Green, H. A. J. (1971). Consumer Theory, MacMillan Press Ltd., London.

Horner, F. B. "Elasticity of the Demand for the Exports of a Single Country," Review of Economics and Statistics 34(1952):326-342.

Intriligator, M. D. (1978). Econometric Models, Techniques and Applications, Prentice-Hall, Englewood Cliffs, New Jersey.

Jarvis, L. S. "Cattle as Capital Goods and Ranchers as Portfolio Managers: An Application to the Argentine Cattle Sector," Journal of Political Economy 82(1974):489-520.

Johnson, P. R., T. Grennes, and M. Thursby. "Trade Models with Differentiated Products," American Journal of Agricultural Economics 61(No. 1, 1979):120-127.

Johnston, J. (1972). Econometric Methods, Second edn. McGraw Hill, New York.

Ospina, E. and C. R. Shumway. "Disaggregated Econometric Analysis of U.S. Slaughter Beef Supply," Technical Monograph 9, Texas Agricultural Experiment Station, Texas A&M University.

Richardson, J. D. "Beyond (But Back to?) The Elasticity of Substitution in International Trade," European Economic Review 4(1973):381-392.

Taplin, J. E. H. "The Elasticity of Demand for the Exports of a Single Country - A Reconsideration," Australian Journal of Agricultural Economics 15(No. 2, 1971):103-108.

Chapter 9

Rod Tyers and Kym Anderson

Imperfect Price Transmission and Implied Trade Elasticities in a Multi-commodity World

In their review of methods for obtaining price elasticities of export demand, Gardiner and Dixit (1987) identify two broad approaches: direct estimation, which involves regressing trade volumes against border prices, and synthetic estimation. Serious econometric difficulties arise with direct estimation methods, as pointed out long ago by Orcutt (1950) and reviewed by Binkley and McKinzie (1981). This has encouraged researchers to consider various synthetic estimation methods.

Ideally, synthetic estimates of trade elasticities for particular commodities should be obtained from a multi-commodity, multi-country dynamic general equilibrium model of the world economy. In the absence of such models, however, analysts have relied on less-complete models of world markets. Often single-commodity models are used, but they miss the important interactions that exist between that commodity and closely related commodities. This paper discusses the trade elasticities implicit in a multi-commodity, multi-country dynamic simulation model of markets for grains, livestock products and sugar (the GLS Model).

The first section of the paper describes the features of the GLS Model. From the viewpoint of obtaining trade elasticities its key features are its multicommodity nature, so that interaction between commodity markets are explicitly incorporated; its dynamic nature, so that both short- and longer-run trade elasticities can be obtained and they can be measured at different points in time; and its inclusion of estimated international-to-domestic price transmission elasticities (short- and longer-run, and differing for producer and consumer prices) so that the stabilizing and insulating behavior of government policy is included endogenously.

The second section of the paper reports the trade elasticities that are implicit in the GLS Model. (The method for obtaining them is detailed in the Appendix.) It shows them for short and long run, with and without market-insulating policies, and for both the early and the late 1980s. The final section summarizes the key findings and suggests some areas for further research.

THE GLS MODEL

The GLS Model is a dynamic, stochastic, multi-commodity simulation model of world food markets. To keep it manageable, it is restricted to the major traded staples, namely, wheat, coarse grain, rice, meat of ruminants (cattle and sheep), meat of nonruminants (pigs and poultry), dairy products and sugar. These seven commodity groups account for about half of world food trade (edible oils and beverages account for most of the rest) and one tenth of global trade in all commodities. Salient features of the model (the algebra of which is detailed in the Appendix) include the following:

- the model is global in coverage, involving 30 countries or country groups spanning the world;
- it incorporates the cross effects, in both production and consumption, between the seven interdependent commodity groups, and separates food and feed demand;
- stockholding behavior is endogenously included, based on empirical analysis of stock level responses to price and quantity changes in each country;
- lags in supply adjustment to price changes are included; and policies in each country are assumed to allow domestic prices to change only gradually (and in some cases not at all) in response to changes in international food prices.

Production behavior is represented by Nerlovian reduced form partial adjustment equations which are linear in the logs of production and producer prices so that constant supply elasticities result. Allowance is also made for the effect on production of land set-aside policies such as those used in the United States. Direct human consumption is characterized by income and price elasticities of demand, which are set to decline over time, while feed consumption by animals is determined by input-output coefficients for each livestock product.

Policies affecting domestic prices are incorporated via econometrically estimated international-to-domestic price transmission equations for each country and commodity. These equations capture both the protection and the stabilization components of food price and trade policies. They are based on estimates of reduced-form Nerlovian partial adjustment equations which distinguish short-run from longer-run elasticities of price transmission. Separate elasticities are used for producer and consumer prices. In general, even the long-run price transmission elasticities are less than unity, reflecting the prevalence of non-tariff protection instruments in food markets. In the face of volatile and declining real prices in international commodity markets, governments limit the extent to which both the long-run trend and the short-run changes in domestic prices follow those of international prices. The smaller the short-run elasticity of price transmission in relation to its long-run counterpart, the greater the degree of market insulation and the more sluggish is the eventual transmission of any sustained change in the international price. In a few extreme cases domestic prices are completely insulated, which means both the short- and long-run elasticities of price transmission are zero (see Appendix Table 9.A.3).

Storage behavior is represented by a combination of risk-neutral competitive speculation and the actions of public agents responding only to quantity triggers. Target levels of closing stocks are set for each commodity at a fixed proportion of trend consumption in importing countries and of trend production in exporting countries. Intertemporal deviations from these target levels depend upon expected speculative profits and the deviation in domestic supply (production plus carry-in) from the long-run trend.

The parameters used in the various equations are based originally on econometric analysis for the period 1961-83 (or for a shorter interval where the full time series was unavailable at the time of analysis) or, in the case of some demand and supply elasticities, on specialized studies in the literature. A sample of elasticities used for Canada, the European Community, Japan and the United States are provided in Table 9.A.2 of the Appendix.

Structurally, the model is simply a set of expressions for quantities consumed, produced and stored, each of which is a function of known past prices and endogenous current prices. The model is solved iteratively by starting from the 1980-82 base period and beginning each subsequent year with the assumption that all prices are the same as those in the preceding year, generating random disturbances in production and calculating new production, consumption and closing stock levels in each country. The resulting excess demands are then totalled and international prices adjusted to move world markets towards clearance. The procedure is then repeated until a satisfactory degree of market clearance has been achieved for each commodity. Thus, the model selects that series of international and domestic prices, production, consumption, and closing stock levels which simultaneously clear all markets in each successive year, from 1983 to 1995. Once 100 simulations of this type have been completed, each using a different set of generated random disturbances from the distributions of each error term, forecast means and standard deviations are calculated for all key variables in the model for each year of the simulation period. They can also be calculated for the base period (1980-82) simply by replacing the values of all time-dependent parameters with their 1980-82 values. The solution procedure is conventional, but it is not based on a standard software package. It is described in more detail for an earlier version of the model in Tyers (1984, 1985) and for the version used here in an appendix of Tyers and Anderson (forthcoming).

IMPLIED TRADE ELASTICITIES IN THE GLS MODEL

It is clear from the above description that trade elasticities are implicit in the GLS Model. They depend on domestic own- and cross-price elasticities of supply and demand in the various countries, on the shares of each country in different markets, and on price transmission elasticities. And they vary according to the time allowed for adjustment, and the year chosen. They are derived by measuring the extent to which the excess demand in the rest of the world for commodity i produced by country K adjusts in response to a change in the price at which country K trades commodity i. (The excess demand of the rest of the world may be a positive or negative quantity, the latter indicating that country K is a net importer of commodity i.) The detailed algebra of this derivation is summarized in equations A.26 to A.43 in the Appendix.

With 30 countries/country groups and seven commodity groups, the total number of own-price trade elasticities is 210 for each year and for each period of adjustment one is interested in. It is therefore necessary to be selective in presenting these parameters. Thus only the very-short-run (same-year), short-run (after one year) and long-run (after full adjustment) trade elasticities are reported here. In Table 1 the own-price elasticities are shown for 1983 for large participants in world food markets wherever those elasticities were less than (+ or -) 40. Table 9.2 shows what those elasticities would have been, according to the GLS Model, if all price transmission elasticities had been unity instead of (typically) less than unity as in the reference case shown in Table 9.1. Table 9.3 then presents the very-short-run elasticities as projected by the model for 1988, for comparison with the 1983 elasticities presented in Tables 9.1 and 9.2. (For details of the policy assumptions used to project beyond 1980-1982, see Anderson and Tyers (1987))

The key points to note from the implied trade elasticities in Table 9.1, for 1983, are as follows:

- the long-run trade elasticities have magnitudes larger than 7.5 for all countries and commodities other than wheat and coarse grain in the United States and wheat in the EC and the USSR;
- the short-run trade elasticities (after one year's adjustment) are never smaller than unity, suggesting that, by reducing the volume of its food imports, no individual economy can increase its export revenue and sustain the increase through a full year or more;
- following the United States the smallest short-run export demand elasticities are faced by Canada, Australia and the EC in the wheat market and by the EC in the market for dairy products;
- the smaller import supply elasticities are faced by the USSR and China in the wheat market and by Japan and the USSR in the market for coarse grain;
- in the rice market, Thailand faces the lowest export demand elasticity, while Japan's small share of the international rice trade results in a comparatively high import demand elasticity at the margin;
- Australia and New Zealand have the lowest export demand elasticities for ruminant meat, followed by Argentina, and the EC and New Zealand have the lowest for dairy products;

- in the sugar market, ten of the countries listed in Table 9.1 face relatively low trade elasticities, with Brazil and the EC facing the lowest on the export side and the USSR facing the lowest on the import side;
- with the exception of coarse grain in the short run, Japan's import supply elasticities are very large, suggesting that Japan has little monopsony power in international food markets.

These trade elasticities would be considerably larger, especially in the short run, if countries did not insulate their domestic markets from changes in international prices. To see this the GLS Model was respecified with all price transmission elasticities set at unity instead of at their econometrically estimated values which are typically less than unity. The implied trade elasticities that emerge with this respecification, summarized in Table 9.2, are in most cases more than twice those in Table 9.1. This result suggests that the insulating component of agricultural policies has a very substantial depressing impact on the size of food trade elasticities and so cannot be ignored in the process of estimating those elasticities, a point stressed by Abbott (1979) and Bredahl, Meyers and Collins (1979). (Removal of the protection component of agricultural policies may or may not raise the food trade elasticities in Table 9.2: it depends on whether production and consumption in the rest of the world is relocated to more- or less-elastic markets, and whether the country in question becomes a larger or smaller participant in the world food economy, as a consequence of the liberalization.)

The trade elasticities in Tables 9.1 and 9.2 refer to 1983. Since then, however, the degree of protection and insulation in world food markets has increased considerably. An attempt has been made to capture those trends in the GLS Model's projection of the world food economy beyond the early 1980s (see Anderson and Tyers (1987) for details). Assuming that projection attempt turns out to be reasonable in retrospect, it allows us to gauge how those implicit trade elasticities may be changing during the 1980s. Table 9.3 shows the very-short-run trade elasticities projected by the model for 1988, for example. In most cases the model's trade elasticities are somewhat smaller for 1988 than for 1983, the only exception being for sugar. And the differences appear to be larger for the case where the stabilization component of policy is removed (price transmission elasticities set at unity). This suggests the degree of insulation as well as the degree of protection has increased during the 1980s: even though the price transmission elasticities are held constant through the projection period (except for US producer prices, which are assumed to be insulated somewhat more from 1986), the changes in the global geographic distribution of food production and consumption have led to the overall degree of market insulation increasing. More generally, the point that is clear from comparing Tables 9.1 and 9.2 with Table 9.3 is that trade elasticities are by no means constant through time.

How do these implicit trade elasticities compare with those adopted in other studies? The most commonly estimated food trade elasticities are the excess demand elasticities facing the United States grain export sector. A sample of such empirical studies over the past decade is summarized in Table 9.4. The GLS Model suggests the elasticities are much higher than those used by Johnson et al. (1985), while our coarse grain elasticities are similar to theirs. Our long-run elasticities are also somewhat larger than those suggested by Bredahl et al. (1979), but they are substantially lower than those suggested by Johnson (1977).

The work by Johnson et al. (1985) is evidently based on national models for the United States developed at the Food and Agricultural Policy Research Institute. These models use estimated trade elasticities to characterize the rest of the world in aggregate. A difficulty with the reliance on such elasticities estimates is that they are invariably based on time series analysis which attempts to explain changes in the level of exports in terms of border price changes. Since exports have other determinants frequently omitted from such studies, specification error leads to the underestimation of the true (ceteris paribus) responsiveness of exports to border price.

In the recent work of Meyers et al. (1987), the elasticities quoted are from a single commodity analysis using an eight-region global model. In that study, trade elasticities are derived by comparing model solutions in which yield functions in the United States have been shifted. The estimates of trade elasticities which result are substantially larger than those from the earlier Johnson et al. and Meyers and Helmar studies. In the case of coarse grain, their one-year elasticity is even larger than our estimate from the GLS Model. The Meyers et al. estimates would, of course have been larger still had the adjustment to a supply shock in one commodity market been permitted to spread across several interacting markets. Although an experiment similar to this is discussed in their article, it was based on simultaneous supply shocks in all markets and hence sheds no light on the (ceteris paribus) trade elasticities which are the subject of this paper.

CONCLUSION

The GLS Model clearly has a number of useful features from the viewpoint of obtaining trade elasticities. The main ones are its multicommodity nature, its inclusion of estimated price transmission elasticities (short- and longer-run, and differing for producer and consumer prices) so that the stabilizing behavior of government policy is included endogenously, and its dynamic nature so that both short- and longer-run trade elasticities can be obtained and they can be measured at different points in time.

Among the desirable features identified by Abbott (1988) that are missing from the GLS Model, however, are that price elasticities are assumed to be the same for upward as for downward price movements, products are assumed to be homogenous and so are indistinguishable by country of origin, and trade is assumed to be competitive or unmanaged. Since the price elasticities of demand and supply in the GLS Model are based on econometric estimates from data which include both upward and downward price movements, they are probably slight underestimates of responses to upward price movements and overestimates of responses to downward price movements. The presence of different types and grades of products within each product group means that if a country trades only a subset of those types or grades, its actual trade elasticities are probably lower than those suggested by the GLS Model in which product homogeneity is assumed. Similarly, the fact that some international trade in these food markets (especially sugar), is managed via State trading and other regulatory devises also means that the GLS Model's implied elasticities are biased upwards.

Bearing in mind the features of the model and its important omissions, the key points to note from the implied trade elasticities reported in this paper are:

- our estimates of short-run (one year response) and long-run trade elasticities facing all individual economies exceed unity, thus ruling out sustained increases in export revenue from restricting excess supply in any individual commodity market;

- except for wheat and coarse grain in the United States and wheat in the USSR, the estimated long-run trade elasticities facing all individual economies have magnitudes larger than 7.5;
- some small economies have low export demand elasticities in certain commodities (rice in Thailand, other grains and beef in Argentina, ruminant meats and dairy products in New Zealand) while large economies do not necessarily have low trade elasticities (Japan, for example);
- trade elasticities would be much larger (in most cases more than double) were it not for the domestic-market-stabilizing behavior of government policies;
- these elasticities vary through time because food production and demands are changing at different rates in countries with different degrees of insulation and because insulation in some is increasing, so that they are smaller in the late 1980s than in the early 1980s; and
- for the United States the grain export demand elasticities obtained from our model appear to be somewhat larger than those from other studies, in part because it allows for more substitution in production and consumption between related products.

Since these synthetic estimates are necessarily less than perfect it is difficult to draw strong conclusions about the precise magnitude of food trade elasticities. They certainly cast doubt, however, on the proposition that any country has strong monopoly or monopsony power in international food markets in anything other than the very short run.

Table 9.1. Excess Demand (Supply) Elasticities Facing Large Participants in World Food Markets, 1980-82*

		Wheat	Coarse grain	Rice	Ruminant meat	Nonruminant meat	Dairy products	Sugar
United States	VSR	-0.6	-0.46	-5.9	(10.0)		-14.2	(6.5)
	SR	-1.2	-1.2	-11.1	(14.6)		-21.2	(8.6)
	LR	-3.3	-3.7	-22.7				(12.2)
European Community (10)	VSR	-3.3	(38.6)			-21.4	-2.5	-4.6
	SR	-7.2				-38.3	-3.6	-6.1
	LR	-18.2					-7.6	-8.8
Japan	VSR	(7.3)	(3.0)	(17.8)	(22.4)		(27.3)	(9.1)
	SR	(15.9)	(9.2)	(33.1)				(12.2)
	LR	(39.7)	(22.6)					(17.6)
Canada	VSR	-1.9	-8.7					(21.6)
	SR	-3.9	-25.3					(28.7)
	LR	-10.3						
Australia	VSR	-3.1	-16.0		-5.6			-7.1
	SR	-6.6			-10.6			-9.4
	LR	-16.9			-35.4			-13.3
New Zealand	VSR				-6.8		-9.9	
	SR				-11.9		-14.8	
	LR				-38.2		-34.3	
USSR	VSR	(1.5)	(3.4)	(34.1)	(12.4)		(23.1)	(2.9)
	SR	(3.3)	(10.4)		(21.8)		(34.9)	(3.8)
	LR	(8.2)	(25.3)					(5.4)
China	VSR	(2.8)		-6.8				(9.9)
	SR	(6.1)		-11.9				(13.0)
	LR	(13.8)		-21.2				(20.6)
Argentina	VSR	-5.2	-5.1		-13.4			-30.0
	SR	-11.5	-15.5		-24.7			
	LR	-30.4	-38.1					
Brazil	VSR	(9.0)			-22.7			-4.5
	SR	(19.8)						-5.9
	LR							-9.2
Thailand	VSR		-19.3	-4.5				-11.5
	SR			-8.3				-15.3
	LR			-17.3				-20.6

*Elasticities not shown have magnitudes greater than 40. Values in parentheses are the excess supply elasticities faced by the country; others are excess demand elasticities. VSR refers to the very short run adjustment within the first year, SR to adjustment after one year and LR to the long run after full adjustment.

Source: Derived from the GLS Model as described in the text.

Table 9.2. Excess Demand (Supply) Elasticities Facing Large Participants in World Food Markets if Price Transmission Elasticities were Unity, 1980-82*

		Wheat	Coarse grain	Rice	Ruminant meat	Nonruminant meat	Dairy products	Sugar
United States	VSR	-3.4	-1.9	-37.1	(36.5)			(27.8)
	LR	-6.9	-6.4					
European Community (10)	VSR	-15.1					-19.3	-16.3
	LR	-31.1					-31.2	-25.1
Japan	VSR	(28.8)	(7.8)					(30.7)
	LR		(31.2)					
Canada	VSR	-8.0	-22.7					
	LR	-17.5						
Australia	VSR	-13.2			-32.8			-23.9
	LR	-28.8						-38.0
New Zealand	VSR				-34.7			
	LR							
USSR	VSR	(5.1)	(8.5)					(9.3)
	LR	(13.0)	(33.3)					(14.9)
China	VSR	(10.0)						(36.4)
	LR	(23.2)						
Argentina	VSR	-23.3	-13.3					
	LR							
Brazil	VSR	(36.5)						-20.3
	LR							-32.7
Thailand	VSR			-27.5				-37.6
	LR			-38.8				

*Elasticities not shown have magnitudes greater than 40. Values in parentheses are the excess supply elasticities faced by the country; others are excess demand elasticities. VSR refers to the very-short-run adjustment within the first year and LR refers to the long run after full adjustment.

Source: Derived from the GLS Model as described in the text, with all international-to-domestic price transmission elasticities set to unity (no insulation).

Table 9.3. Very-Short-Run Excess Demand (Supply) Elasticities Facing Large Participants in World Food Markets, Projected to 1988*

		Wheat	Coarse grain	Rice	Ruminant meat	Nonruminant meat	Dairy products	Sugar
United States	ref.	-0.51	-0.27	-4.7			-4.0	(13.2)
	PTE=1	-2.5	-1.1	-29.1			-27.5	
European Community (10)	ref.	-2.1			-11.3	-9.2	-1.7	-3.4
	PTE=1	-7.3				-24.3	-12.1	-10.7
Japan	ref.	(8.2)	(2.8)	(20.7)	(25.3)		(31.0)	(9.9)
	PTE=1	(31.4)	(7.0)					(33.7)
Canada	ref.	-1.8	-7.0					(24.1)
	PTE=1	-7.1	-17.5					
Australia	ref.	-2.9	-13.7		-5.1			-7.5
	PTE=1	-11.6	-33.9		-28.3			-25.4
New Zealand	ref.				-6.4		-11.2	
	PTE=1						-31.5	
USSR	ref.	(2.9)	(5.6)	(33.6)			-9.3	(3.3)
	PTE=1	(9.1)	(13.3)					(10.9)
China	ref.	(2.5)	(8.6)	-3.9			(28.3)	(7.2)
	PTE=1	(9.1)	(19.0)	-14.6				(22.2)
Argentina	ref.	-5.0	-4.7		-12.7			-26.1
	PTE=1	-20.3	-11.8					
Brazil	ref.	(9.0)			-24.0			-3.6
	PTE=1	(35.5)						-14.9
Thailand	ref.		-18.7	-3.6				-13.8
	PTE=1			-22.0				

*Elasticities not shown have magnitudes greater than 40. Values in parentheses are the very-short-run excess supply elasticities faced by the country; others are very-short-run excess demand elasticities. Ref. refers to the reference scenario (corresponding with Table 1 for the base period), PTE=1 refers to the scenario in which all price transmission elasticities are set at unity (corresponding with Table 9.2 for the base period).

Source: Derived from the GLS Model as described in the text.

Table 9.4. Comparison of Excess Demand Elasticities Facing U.S. Grain Exporters*

		This study (from Tables 1 & 3) 1983	1988	Johnson* (1977, Table 1)	Bredahl at al. (1979, Table 4)	Johnson et al. (1985, Table 1)	Meyers and Helmar (1986, Table 5)	Other studies surveyed by Gardiner and Dixit (1987, Tables 2, 3, and 5)	Meyers et al. (1987 Tables 1, 2)
Wheat	-very short run	-0.60	-0.51						
	-short run	-1.20	-1.00			-0.16	-0.11	-0.14 to -3.1	-0.90
	-long run	-3.30	-2.90	-6.7	-0.00 to -1.7	"near 1.0"		-0.23 to -5.0	-1.27
Coarse grain	-very short run	-0.46	-0.27						
	-short run	-1.20	-0.74			-0.16	-0.29	-0.30 to -1.5	-1.38
	-long run	-3.70	-2.30	-10.2	-0.09 to -1.3**	"near 1.0"		-0.90 to -3.3	-1.59
Rice	-very short run	-5.90	-4.70						
	-short run	-11.10	-8.90			-0.25		-0.46 to -0.68	
	-long run	-22.70	-18.10			"near 1.0"		-7.00	

*Assuming price transmission elasticities are unity.
**For corn only; -0.29 to -2.6 for sorghum.

APPENDIX

THE GLS MODEL AND DERIVATION OF ITS IMPLIED TRADE ELASTICITIES

This appendix first details the main behavioral equations of the GLS Model described in the paper, which are summarized in Table 9.A.1 below. It then explains how the trade elasticities implicit in the model are derived. The Appendix concludes with a summary of the main domestic demand and supply elasticities in the model for Canada, the European Community, Japan and the United States, and with the international-to-domestic price transmission elasticities in Table 9.A.3. (See Tyers and Anderson (forthcoming) for the complete parameter set for all 30 countries and country groups in the model.)

The Behavioral Equations

Production behavior is represented by a "partial adjustment" model which is linear in the logs of production and producer prices (equations A.1 and A.2 in Table 9.A.1 below). Allowance is made for the effect on production of land set-aside policies such as those presently active in the United States (equation A.3). Production in each country and country group is subject to random disturbances to equation (A.2) which take the form indicated in equation A.4.

Direct consumption is assumed to be characterized by constant income and price elasticities of demand which decline over time (equations A.7 and A.8). The demand for livestock feeds, however, is based on a fixed input-output coefficient for each livestock commodity (equation A.9). This coefficient is assumed to change with time as the proportion of all livestock output in the country which is feed-based increases. Because short-run changes in livestock production do not generally follow the corresponding changes in the livestock populations to be fed, a steady-state level of livestock output is identified which corresponds more closely in the short run to animal feed requirements. This steady-state output is a weighted moving average of past levels of output (equation A.10).

Storage behavior is represented by a combination of risk-neutral competitive speculation and the actions of public agents responding only to quantity triggers. The parameters are derived empirically using time series analysis and equation (A.14). Target levels of closing stocks are set for each commodity at a fixed proportion of trend consumption in importing countries and of trend production in exporting countries (equation A.15). Intertemporal deviations from the target levels depend upon expected speculative profits and the deviation in domestic supply (production plus carry-in) from its long-run trend.

Policies affecting trade are incorporated through price transmission equations for each country and commodity. These equations cover both the protection and the market insulation components of food price and trade policies. They express domestic producer and consumer prices in each country in terms of "world indicator prices," P_{it}.

Actual border prices facing each country in the model differ from these indicator prices by a factor reflecting differences between countries in the quality of the commodities they trade and the different costs of freight and insurance associated with shipment to or from major trading partners. These factors are assumed to remain constant, even where changes of trade direction take place (equation A.22).

The price transmission equations assume reduced-form Nerlovian lagged adjustment, with short- and long-run elasticities of price transmission distinguished. Note that the determination of producer prices are based on parameters which generally differ from those determining consumer prices. In this model, the long-run elasticity of price transmission is generally less than unity, reflecting the prevalence of non-tariff protection instruments in food markets. Unless markets are specified as completely insulated (both the short- and long-run elasticities of price transmission are zero), governments facing volatile international commodity markets limit both the extent to which the trend of domestic prices can follow that of international prices and, in the interests of domestic price stability, the short-run movements in domestic prices which would otherwise occur as world prices fluctuate. The smaller the short-run elasticity of price transmission in relation to its long-run counterpart, the greater the degree of market insulation and the more sluggish is the eventual transmission of any sustained change in the international price.

Analytical Derivation of Implied Trade Elasticities

The GLS Model treats each of its 30 countries or country groups as "large," in that changes in the trade volumes of each induce non-zero changes in international prices. Of special interest are the excess supply or demand elasticities (of "the rest of the world") which face each country. Where these are comparatively small in magnitude for particular countries, those countries have the capacity to increase domestic welfare at the expense of the rest of the world by imposing trade taxes.

Consider a world comprising N countries (K = 1, ..., N) and several commodities. Let m_i^K be the excess demand in the rest of the world for commodity i produced by country K. This excess demand might be either positive or negative, the latter indicating that country K is a net importer. The excess demand elasticity is then the proportional adjustment which occurs in m_i^K in response to a given proportional change in the price at which country K trades commodity j (where j may or may not equal i).

$$e_{ij}^K = \frac{dm_i^K}{m_i^K} / \frac{dP_j}{P_j} \tag{A.26}$$

We present here a reliable analytical approximation to the elements of the matrix $[e_{ij}^K]$, based on the structural equations and parameters of the GLS Model. First, we seek an expression for the adjustment in excess demand in the rest of the world, dm_i^K, in terms of changes in prices. This adjustment is simply the sum of the adjustments made in the countries, k, of the rest of the world group.

$$dm_i^K = \sum_{k \neq K} \left\{ dc_{ik}^D + dc_{ik}^F - dq_{ik} + ds_{ik} \right\} \tag{A.27}$$

The first term is the adjustment in direct demand, the second is that in feed demand, the third is that in production and the final term is the adjustment in stock levels. It is convenient to express each of these terms as functions of domestic prices in country k (≠K). The direct consumption adjustment is readily derived from the differentiation of equation (A.7) as

$$dc_{ik}^{D} = c_{ik}^{D} \sum_{j} a_{ijk} \frac{dp_{jk}^{C}}{p_{jk}^{C}} \tag{A.28}$$

The corresponding adjustments in feed demand, production and stock levels depend on the length of run. Here we consider three extreme cases discussed in the paper: adjustment in the same year, after one year, and in the long run.

Adjustment in the same year:

The relationship between feed demand and livestock production is based in the model on fixed input-output coefficients which link that demand to "steady-state" livestock output, as is specified in equations (A.9) and (A.10). Taking derivatives of the terms in these equations and in (A.1) and (A.2), which depend on prices in the current year, and summing over the feed-consuming sectors, n, we obtain the following expression:

$$dc_{ik}^{F} = \sum_{n} \alpha_{in} \beta_{nk} \tau_{onk} q_{nk} \delta_{nk} \sum_{j} b_{onjk} \frac{dp_{jk}}{p_{jk}} \tag{A.29}$$

The prices in this equation are consumer prices where commodity j is a livestock feed and producer prices where it is a livestock product (see equation A.5).

The corresponding production adjustment is derived in the same way, from equations (A.1) and (A.2):

$$dq_{ik} = q_{ik}\delta_{ik}\sum_j b_{oijk}\frac{dp_{jk}}{p_{jk}} \tag{A.30}$$

where the same rule applies regarding the prices in the final term (equation A.5).

The stock level adjustment requires some manipulation. The quantity held in storage is given by equation (A.14). This equation has the level of stocks on both sides, however, since this level affects the physical cost of storage and hence the profitability of holding stocks. We first correct this

$$s_{ikt}\left(1 + \frac{\pi_{ik}\theta_{ik}z_{ikt}}{\bar{s}_{ikt}}\right) = \pi_{ik}z_{ikt}\left[p^S_{ikt+1} - (1+r_k)\,P^S_{ikt}\right]$$

$$+ \psi_{ik}\left[q_{ikt} + s_{ikt-1} - \bar{q}_{ikt} - \bar{s}_{ikt}\right] + w_{ik}z_{ikt} \tag{A.31}$$

In practice, the prices facing stock-holders, P^S_{ikt}, are in some instances domestic consumer prices and in others border prices. To simplify exposition, we consider only the case where $P^S_{ikt} = P^C_{ikt}$. Since our aim is to approximate the global excess demand elasticities, it is convenient to ignore the indirect dependence of z_{ikt}, $\bar{q}_{ikt}$ and $\bar{s}_{ikt}$ on current period prices. Thus, noting equation (A.17) and again dropping the time subscript, we have the approximate total derivative:

$$ds_{ik} \approx \left(\frac{\bar{s}_{ik}}{\bar{s}_{ik} + \pi_{ik}\theta_{ik}z_{ik}}\right)\left[\psi_{ik}dq_{ik} - \pi_{ik}z_{ik}\left(\frac{3}{4} + r_k\right)dp^C_{ik}\right] \tag{A.32}$$

Replacing, dp^C_{ik} with $\bar{p}_{ik0}\left(dp^C_{ik}/p^C_{ik}\right)$ and substituting for dq_{ik} from equation (A.30), equation (A.32) is reduced to an expression in the proportional changes in producer and consumer prices. If we then set c^D_{ik}, q_{ik}, z_{ik}($\bar{c}_{ik}$ or $\bar{q}_{ik}$) equal to the appropriate quantity shifters and $\bar{s}_{ik}$ to w_{ik} z_{ik} (the steady-state storage level), equations (A.28), (A.29), (A.30) and (A.32) can be combined in equation (A.26) to yield the following expression for the adjustment in the overall excess demand of country k in terms of price changes:

$$dm_{ik} \approx \sum_j \left[A_{jk} \frac{dp^C_{jk}}{p^C_{jk}} + B_{jk} \frac{dp^P_{jk}}{p^P_{jk}} \right] \tag{A.33}$$

where A_{jk} and B_{jk} are constants in any given year.

An expression in international prices is needed, however. This is achieved by noting, from equations (A.20) and (A.21)

$$\frac{dp^C_{jk}}{p^C_{jk}} = \phi^{CSR}_{jk} \frac{dP_j}{P_j}, \quad \frac{dp^P_{jk}}{p^P_{jk}} = \phi^{PSR}_{jk} \frac{dP_j}{P_j} \tag{A.34}$$

where ϕ^{CSR}_{jk} and ϕ^{PSR}_{jk} are the same-year price transmission elasticities. The adjustment in the excess demand of "the rest of the world" for the exports of focus country, K, is then:

$$dm^K_i = \sum_j \frac{dP_j}{P_j} \sum_{k \neq K} \left(\phi^{CSR}_{jk} A_{jk} + \phi^{PSR}_{jk} B_{jk} \right) \tag{A.35}$$

where the sum over countries k excludes the focus country, K. From equation (A.26), the excess demand elasticity facing country K is then:

$$e_{ij}^{K} = \frac{\sum_{k \neq K} \left(\phi_{jk}^{CSR} A_{jk} + \phi_{jk}^{PSR} B_{jk} \right)}{m_{ik}} \tag{A.36}$$

Adjustment after one year:

In this case the terms in equation (A.27) depend on the sum of the same-year and the one-year lagged adjustments in consumption, production and stocks. It is convenient to begin with the production adjustment. For equations (A.1) and (A.2) it is possible to eliminate target production and assemble a reduced form expression for q_{ik} in terms of lagged production and prices. From repeated applications of this reduced form it is evident that the one-year production response is:

$$dq_{ik}^{1} = q_{ik} \delta_{ik} \sum_{j} \left[(2-\delta_i) b_{0ijk} + b_{1ijk} \right] \frac{dp_{jk}}{p_{jk}} \tag{A.37}$$

where the prices are again those determined in equation (A.5).

The response of feed demand after one year then depends on the adjustment of steady state livestock production to changes in price. In this case two terms in equation (A.11) are active and the feed consumption adjustment is:

$$dc_{ik}^{F} = \sum_{n} \alpha_{in} \beta_{nk} \left(\tau_{0nk} dq_{nk}^{0} + \tau_{1nk} dq_{nk}^{1} \right) \tag{A.38}$$

where dq_{nk}^{0} is the same-year production adjustment, given in equation (A.30), and dq_{nk}^{1} is the one-year adjustment, given in equation (A.37).

The one-year adjustment can be derived from the application of a price change to equation (A.31) in successive time periods. Similar approximations to those underlying equation (A.32) are required. The result is:

$$ds_{ik} \approx \lambda_{ik}\psi_{ik}dq^1_{ik} - \lambda_{ik}\pi_{ik}Z_{ik}\left[r_k + \frac{1}{2} + \lambda_{ik}\psi_{ik}\left(r_k + \frac{3}{4}\right)\right]\bar{p}_{iko}\frac{dp_{ik}}{p_{ik}}$$

(A.39)

where, again, the price variable is either a domestic consumer price or a border price, depending which performed best in the econometric analysis. As before, we assume for expository purposes that the domestic consumer price is appropriate.

The adjustment in country k's overall excess demand can then be formulated in terms of changes in domestic producer and consumer prices, as in equation (A.33). It remains to link these domestic price changes to changes in international prices. This is readily achieved by applying changes in international prices to equations (A.20) and (A.21) in successive years. The result is the formulation of the one-year price transmission elasticity in terms of the corresponding same-year and long-run elasticities, which are both parameters in the model. Since the expression is identical for domestic consumer and producer prices, we drop the P and C superscripts:

$$\frac{dp_{jk}}{p_{jk}} = \phi^{SR}_{jk}\left(2 - \frac{\phi^{SR}_{jk}}{\phi^{LR}_{jk}}\right)\frac{dP_j}{P_j} \qquad \text{(A.40)}$$

Adjustment in the long run:

In this case, the global excess demand elasticities embody all long-run quantity adjustments to price changes. Again, equations (A.26) and (A.27) apply, except that changes in stock levels in the long run do not effect annual excess demand. Accordingly, the third term in (A.26) is dropped. The changes in direct and feed consumption are now:

$$dc_{ik}^{D} = c_{ik}^{D} \sum_{j} a_{ijk} \frac{dp_{jk}^{C}}{p_{jk}^{C}} \tag{A.41}$$

$$dc_{ik}^{F} = \sum_{n} \alpha_{in} \beta_{nk} q_{nk} \sum_{j} \left(b_{0njk} + b_{1njk} + b_{2njk} \right) \frac{dp_{jk}}{p_{jk}} \tag{A.42}$$

Note that long-run supply elasticities are used in the second term - the sum of the lagged supply response elasticities of equation (A.1). These same elasticities are the basis for the change in production:

$$dq_{ik} = q_{ik} \sum_{j} \left(b_{0ijk} + b_{1ijk} + b_{2ijk} \right) \frac{dp_{jk}}{p_{jk}} \tag{A.43}$$

Substituting (A.41), (A.42) and (A.43) into (A.26), and setting dsik to zero, we derive an expression like equation (A.33). The expression for the long-run demand elasticity then differs from equation (A.36) only in that the constants A_{jk} and B_{jk} differ and the price transmission elasticities which appear are the long run values, ϕ_{jk}^{CLR} and ϕ_{jk}^{PLR}.

Table 9.A.1: The GLS Model's Behavioral Equations

Production

Target output level:

$$q^*_{ikt} = q^T_{ikt} \prod_j \left\{ \left(\frac{p_{jkt}}{\bar{p}_{jk0}}\right)^{b_{0ijk}} \left(\frac{p_{jkt-1}}{\bar{p}_{jk0}}\right)^{b_{1ijk}} \left(\frac{p_{jkt-2}}{\bar{p}_{jk0}}\right)^{b_{2jk}} \right\} \tag{A.1}$$

Partial adjustment to target output level:

$$q_{ikt} = q^T_{ikt} \left(\frac{q_{ikt-1}}{q^T_{ikt-1}}\right) \left(\frac{q^*_{ikt}}{q^T_{ikt}} \Big/ \frac{q_{ikt-1}}{q^T_{ikt-1}}\right)^{\delta_{ik}} e^{\varepsilon_{ikt}} \tag{A.2}$$

Production shifter:

$$q^T_{ikt} = q^T_{ik0}\,(1 - \lambda_{ikt})\, e^{g_{ik}t} \tag{A.3}$$

where λ_{ikt} is the fraction by which output is reduced by set-aside policy and g_{ik} is the growth rate due to cost-reducing technical change.

Random production disturbances:

$$\varepsilon_{kt} \sim N(0, U_k) \tag{A.4}$$

Table 9.A.1. Continued.

Prices faced by producers:

$$P_{jkt} = \begin{cases} P^{P}_{jkt} & \text{where j is a production substitute.} \\ P^{C}_{jkt} & \text{where i is a livestock product and j is an animal feed.} \end{cases} \tag{A.5}$$

Consumption

Total consumption:

$$c_{ikt} = c^{D}_{ikt} + c^{F}_{ikt} \tag{A.6}$$

Direct human consumption:

$$c^{D}_{ikt} = c^{TD}_{ikt} \prod_{j} \left(\frac{p^{C}_{ikt}}{\bar{p}^{C}_{jk0}} \right)^{a_{ijk}} \tag{A.7}$$

Direct consumption shifter:

$$c^{TD}_{ikt} = c^{TD}_{ik0} \left(\frac{N_{kt}}{N_{k0}} \right) \left(\frac{y_{kt}}{y_{k0}} \right)^{n_{ik}} \tag{A.8}$$

Table 9.A.1. Continued.

Consumption as animal feed:

$$c^F_{jkt} = \sum_i \alpha_{ji} \beta_{ikt} q^S_{ikt} \tag{A.9}$$

Steady-state livestock output at year t animal population:

$$\bar{q}^S_{ikt} = \bar{q}_{ikt} \left[1 + \tau_{0ik} \left(\frac{q_{ikt}}{\bar{q}_{ikt}} - 1 \right) + \tau_{1ik} \left(\frac{q_{ikt-1}}{\bar{q}_{ikt}} \frac{q^T_{ikt}}{q^T_{ikt-1}} - 1 \right) + \tau_{2ik} \left(\frac{q_{ikt-2}}{\bar{q}_{ikt}} \frac{q^T_{ikt}}{q^T_{ikt-2}} - 1 \right) \right] \tag{A.10}$$

$$\bar{q}_{ikt} = \frac{1}{3} \left(q_{ikt} + q_{ikt-1} \frac{q^T_{ikt}}{q^T_{ikt-1}} + q_{ikt-2} \frac{q^T_{ikt}}{q^T_{ikt-2}} \right) \tag{A.11}$$

Fraction of the total supply response after v years:

$$\tau_{vik} = \frac{b_{vijk}}{b_{0ijk} + b_{1ijk} + b_{2ijk}} \tag{A.12}$$

where commodity j is coarse grain (the one commodity which is both directly consumed and fed to animals).

Table 9.A.1. Continued.

Total consumption shifter:

$$c^{T}_{jkt} = c^{TD}_{jkt} + \sum_{i} \alpha_{ji}\beta_{ikt}q^{T}_{ikt} \tag{A.13}$$

Closing Stocks

$$\frac{s_{ikt}}{z_{ikt}} = \pi_{ik}\left[p^{S}_{ikt+1} - (1+r_k)p^{S}_{ikt} - \theta_{ik}\frac{s_{ikt}}{\bar{s}_{ikt}}\right] + \Omega_{ik}\frac{q_{ikt} + s_{ikt-1} - \bar{q}_{ikt} - \bar{s}_{ikt}}{z_{ikt}} + w_{ik} \tag{A.14}$$

Quantity shifter:

$$z_{ikt} = \begin{cases} \bar{q}_{ikt} & \text{if} \quad \bar{q}_{ikt} > \bar{c}_{ikt} \\ \bar{c}_{ikt} & \text{if} \quad \bar{q}_{ikt} < \bar{c}_{ikt} \end{cases} \tag{A.15}$$

Consumption moving average:

$$\bar{c}_{ikt} = \frac{1}{3}\left(c_{ikt} + c_{ikt-1}\frac{c^{T}_{ikt}}{c^{T}_{ikt-1}} + c_{ikt-2}\frac{c^{T}_{ikt}}{c^{T}_{ikt-2}} \right) \tag{A.16}$$

Expected stock-holder price in t+1:

$$P^{S}_{ikt+1} = \frac{1}{4}\left(P^{S}_{ikt} + P^{S}_{ikt-1} + P^{S}_{ikt-2} + P^{S}_{ikt-3} \right) \tag{A.17}$$

Table 9.A.1. Continued.

Stock-holder price:

$$P^{S}_{ikt} = \begin{cases} p^{C}_{ikt} & \text{where stocks traded domestically.} \\ P^{B}_{ikt} & \text{where stocks traded at the border.} \end{cases} \tag{A.18}$$

Stock-level moving average:

$$\bar{s}_{ikt} = \frac{1}{3}\left(s_{ikt} + s_{ikt-1}\frac{z_{ikt}}{z_{ikt-1}} + s_{ikt-2}\frac{z_{ikt}}{z_{ikt-2}} \right) \tag{A.19}$$

Price Transmission

Domestic consumer prices:

$$p^{C}_{ikt} = \rho^{C}_{ikt}\,\bar{P}^{B}_{ik0}\left(\frac{p^{C}_{ikt-1}}{\rho^{C}_{ikt-1}\,\bar{P}^{B}_{ik0}} \right)^{\left(1 - \frac{\phi^{CSR}_{ikt}}{\phi^{CLR}_{ikt}}\right)} \left(\frac{P^{B}_{ikt}}{\bar{P}^{B}_{ik0}} \right)^{\phi^{CSR}_{ikt}} \tag{A.20}$$

Domestic producer prices:

$$p^{P}_{ikt} = \rho^{P}_{ikt}\,\bar{P}^{B}_{ik0}\left(\frac{p^{P}_{ikt-1}}{\rho^{P}_{ikt-1}\,\bar{P}^{B}_{ik0}} \right)^{\left(1 - \frac{\phi^{PSR}_{ikt}}{\phi^{PLR}_{ikt}}\right)} \left(\frac{P^{B}_{ikt}}{\bar{P}^{B}_{ik0}} \right)^{\phi^{PSR}_{ikt}} \tag{A.21}$$

Table 9.A.1. Continued.

Border price:

in year t

$$P^{B}_{ikt} = h_{ik}P_{it} / x_{kt} \qquad \text{(A.22)}$$

in base period

$$\overline{P}^{B}_{ik0} = h_{ik}\overline{P}_{i0} / \overline{x}_{k0} \qquad \text{(A.23)}$$

Excess Demand

$$m_{ikt} = c_{ikt} + s_{ikt} - q_{ikt} - s_{ikt-1} \qquad \text{(A.24)}$$

Global Market Clearing Condition

$$\sum_{k}^{n} m_{ikt} = 0 \qquad \text{(A.25)}$$

Nomenclature

Indices:

i,j	Commodity counters.
k	Counter for countries and country-group
v	Counter for years of lag in production response.
t	Time.

Table 9.A.1. Continued.

Quantities:

q_{ikt}	Production of commodity i in country k and year t.
q^T_{ikt}	Production shifter. This would be the trend of production in the absence of producer price changes.
q^S_{ikt}	Livestock output which would take place in a steady state with the livestock numbers actually prevailing in year t.
c_{ikt}	Consumption of commodity i in country k and year t.
c^D_{ikt}	Direct consumption.
c^F_{ikt}	Consumption as animal feed.
c^{TD}_{ikt}	Direct consumption shifter. This would be the trend of direct consumption in the absence of consumer price changes.
c^T_{ikt}	Total consumption shifter.
m_{ikt}	Excess demand for commodity i of country k in year t.
s_{ikt}	Closing stock of comodity i in country k in year t.
e_{ikt}	Proportional randum production disturbance.
U_k	Variance-covariance matrix of random production disturbances across commodities in country k.

Table 9.A.1. Continued.

Prices:

p_{ikt} Domestic price of commodity i in country k and year t. This price is used in equation (A.1). Equation (A.5) indicated where a producer or consumer price is implied.

$\bar{p}_{ik0}$ Average domestic price in the base period, 1980-82, to correspond with p_{ikt}.

p^{s}_{ikt} Price at which stocks are traded. Equation (A.20) indicates where this is the domestic consumer price and where it is the border price.

P_{it} International indicator price-standard trading price of commodity i in year t. Actual border prices differ from this as per equation (A.28).

$\bar{P}_{io}$ Standard international price in the base period, 1980-82. Based on f.o.b. export prices at the major ports in Thailand (rice), Canada (wheat), USA (maize, pork and poultry), Australia (beef), New Zealand (milk) and the Carribean (sugar).

Parameters:

b_{vijk} The elasticity of target production of commodity i with respect to the price of j in country k. The subscript v indicates the length of response lag: v=0: less than one year; v=1: one year; v=2: two years.

δ_{ik} The partial adjustment elasticity for the production of commodity i in country k.

g_{ik} The growth rate in the trend of production of commodity i in country k which would be sustained with constant real producer prices.

Table 9.A.1. Continued.

a_{ijk}	The elasticity of direct demand for commodity i with respect to the consumer price of j.
η_{ik}	The income elasticity of demand for commodity i in country k.
α_{ij}	Input-output coefficient: the quantity of commodity i used in the production of one unit of commodity j. This is assumed to be invariant across countries in the grain-fed components of livestock sectors.
β_{ikt}	The fraction of the production of livestock product i which is grain-fed in country k and year t.
τ_{vik}	For livestock products, this parameter measures the fraction of the total target supply response which occurs v years after the price change.
r_k	The real rate of interest in country k.
θ_{ik}	Steady-state marginal cost of storage of commodity i in country k.
π_{ik}	The response of closing stocks of commodity i to expected speculative storage profits in country k.
Ω_{ik}	The response of closing stocks to the carry-in level of commodity in country k.
w_{ik}	The steady-state level of working stocks of commodity i, as a proportion of trend consumption in importing countries and of production in exporting countries.

Table 9.A.1. Continued.

h_{ik} The base-period ratio of the appropriate border price of commodity i in country k, $\bar{P}^{B}_{ik0}$ to the standard or indicator international price, $\bar{P}^{B}_{i0}$. These divergences reflect quality differences, freight costs and the pattern of concessional sales, all of which are assumed to remain constant.

Exogenously projected variables:

N_{kt} The population of country k in year t.

y_{kt} National income per capita in country k in year t.

x_{kt} Real exchange rate in US$ per local currency unit of country k, in year t.

$\bar{x}_{KO}$ Base period average real exchange rate.

Policy parameters:

ρ^{P}_{ikt} The target nominal protection coefficient for producers of commodity i in country k. This can be changed through time, hence the time subscript. From equations (A.24) through (A.27) it is clear that the actual nominal protection coefficient fluctuates around this value, depending on the degree of market insulation.

ρ^{C}_{ikt} The corresponding ratio indicating the degree to which consumer prices are distorted by government policy.

ϕ^{PSR}_{ikt} The short-run elasticity of transmission of international market changes to domestic producer prices of

Table 9.A.1. Continued.

	commodity i in country k. This measures the degree of insulation provided by policy. Since this can change through time, the parameter may also be adjusted along some exogenous time-path.
ϕ^{PLR}_{ikt}	The corresponding long-run elasticity of price transmission for producer prices of commodity i in country k. This may be less than unity where distortions depend on non-tariff barriers.
ϕ^{CSR}_{ikt}	The short-run elasticity of price transmission for commodity i in country k.
ϕ^{CLR}_{ikt}	The corresponding long-run elasticity of price transmission for consumer prices of commodity i in country k.
ikt	The fraction by which set-aside policies shift the supply curve of commodity i in country k to the left.

Table 9.A.2. Key Demand, Supply and Price Transmission Elasticities for Major Industrial Countries

	Reference consumption (kt)	Elasticity of demand with respect to the price of: Rice	Wheat	C. Grain	Sugar	Dairy	R. Meat	NR. Meat
Canada								
Rice	107	-0.30	0.10	0.10	0.0	0.0	0.0	0.0
Wheat	5505	0.0	-0.18	0.05	0.0	0.0	0.0	0.0
C. Grain	17075	0.0	0.15	-0.20	0.0	0.0	0.0	0.0
Sugar	992	0.0	0.0	0.02	-0.08	0.0	0.0	0.0
Dairy	6999	0.0	0.0	0.0	0.0	-0.40	0.0	0.0
R. Meat	1099	0.0	0.0	0.0	0.0	0.0	-0.65	0.30
NR. Meat	1285	0.0	0.0	0.0	0.0	0.0	0.25	-0.75
Indirect Demand Parameters for Coarse Grain:								
Shares of livestock sectors grain-fed						0.78	0.78	1.00
Grain use per unit of output						0.40	6.00	5.00
The European Community								
Rice	945	-0.80	0.25	0.10	0.0	0.0	0.0	0.0
Wheat	47850	0.01	-0.30	0.02	0.0	0.0	0.0	0.0
C. Grain	70195	0.0	0.17	-0.20	0.05	0.0	0.0	0.0
Sugar	10533	0.0	0.0	0.01	-0.12	0.0	0.0	0.0
Dairy	107187	0.0	0.0	0.0	0.0	-0.40	0.02	0.02
R. Meat	7632	0.0	0.0	0.0	0.0	0.02	-0.60	0.25
NR. Meat	14029	0.0	0.0	0.0	0.0	0.02	0.26	-0.90
Indirect Demand Parameters for Coarse Grain:								
Shares of livestock sectors grain-fed						0.38	0.38	0.38
Grain use per unit of output						0.40	6.00	5.00
Japan								
Rice	10472	-0.23	0.03	0.01	0.0	0.0	0.0	0.0
Wheat	6331	0.24	-0.60	0.14	0.0	0.0	0.0	0.0
C. Grain	19436	0.16	0.25	-0.40	0.0	0.0	0.0	0.0
Sugar	2851	0.01	0.0	0.0	-0.05	0.0	0.0	0.0
Dairy	8113	0.0	0.0	0.0	0.0	-0.80	0.0	0.0
R. Meat	706	0.0	0.0	0.0	0.0	0.0	-1.40	0.40
NR. Meat	2904	0.0	0.0	0.0	0.0	0.0	0.25	-1.00
Indirect Demand Parameters for Coarse Grain:								
Shares of livestock sectors grain-fed						0.46	0.46	1.00
Grain use per unit of output						0.40	6.00	5.00
The United States								
Rice	2015	-0.20	0.08	0.04	0.0	0.0	0.0	0.0
Wheat	26958	0.01	-0.12	0.06	0.0	0.0	0.0	0.0
C. Grain	155456	0.01	0.08	-0.20	0.07	0.0	0.0	0.0
Sugar	8693	0.0	0.0	0.05	-0.20	0.0	0.0	0.0
Dairy	60503	0.0	0.0	0.0	0.0	-0.30	0.02	0.01
R. Meat	11190	0.0	0.0	0.0	0.0	0.02	-0.50	0.20
NR. Meat	13825	0.0	0.0	0.0	0.0	0.01	0.20	-0.80
Indirect Demand Parameters for Coarse Grain:								
Shares of livestock sectors grain-fed						0.67	0.67	1.00
Grain use per unit of output						0.40	6.00	5.00

Table 9.A.2. Continued.

	Reference consumption (kt)	Long-run elasticity of supply with respect to the price of: Rice	Wheat	C. Grain	Sugar	Dairy	R. Meat	NR. Meat
Canada								
Rice	0	0.0	0.0	0.0	0.0	0.0	0.0	0.0
Wheat	26042	0.0	0.53	-0.22	0.0	0.0	-0.60	0.0
C. Grain	23130	0.0	-0.34	0.68	0.0	0.0	0.0	0.0
Sugar	132	0.0	0.0	0.0	0.50	0.0	0.0	0.0
Dairy	7772	0.0	0.0	-0.10	0.0	0.50	0.0	-0.08
R. Meat	1092	0.0	0.0	-0.28	0.0	0.08	0.60	-0.18
NR. Meat	1406	0.0	0.0	-0.25	0.0	-0.09	-0.14	0.89
The European Community								
Rice	699	0.40	0.0	0.0	0.0	0.0	0.0	0.0
Wheat	57772	0.0	0.90	-0.66	-0.06	0.0	0.0	0.0
C. Grain	67299	0.0	-0.51	0.92	-0.05	0.0	0.0	0.0
Sugar	14164	0.0	-0.10	-0.10	0.50	0.0	0.0	0.0
Dairy	118757	0.0	0.0	-0.01	0.0	0.51	-0.03	0.0
R. Meat	7520	0.0	0.0	-0.01	0.0	0.12	1.02	-0.48
NR. Meat	14813	0.0	0.0	-0.37	0.0	0.0	-0.30	1.14
Japan								
Rice	9375	0.20	0.0	0.0	0.0	0.0	0.0	0.0
Wheat	675	0.0	0.60	-0.30	0.0	0.0	0.0	0.0
C. Grain	399	0.0	-0.40	0.60	0.0	0.0	0.0	0.0
Sugar	853	0.0	0.0	0.0	0.50	0.0	0.0	0.0
Dairy	6798	0.0	0.0	-0.06	0.0	0.80	-0.09	0.0
R. Meat	478	0.0	0.0	-0.06	0.0	0.04	0.80	-0.10
NR. Meat	2619	0.0	0.0	-0.23	0.0	0.0	-0.06	0.99
The United States								
Rice	4713	0.75	-0.20	0.0	-0.04	0.0	0.0	0.0
Wheat	72301	-0.04	0.80	-0.53	0.0	0.0	0.0	0.0
C. Grain	211494	0.0	-0.28	0.75	0.0	0.0	0.0	0.0
Sugar	5321	-0.04	0.0	0.0	0.28	0.0	0.0	0.0
Dairy	61807	0.0	0.0	-0.08	0.0	0.85	-0.20	0.0
R. Meat	10578	0.0	0.0	-0.24	0.0	0.03	0.72	-0.16
NR. Meat	13991	0.0	0.0	-0.38	0.0	0.0	-0.13	1.12

Table 9.A2. Continued

Short-run elasticity of supply with respect to the price of:

	Rice	Wheat	Coarse Grain			Sugar	Dairy			Ruminant meat			Nonruminant meat		
	t-1	t-1	t	t-1	t-2	t-1	t	t-1	t-2	t	t-1	t-2	t	t-1	t-2
Canada															
Rice	0.0	0.0		0.0		0.0		0.0			0.0			0.0	
Wheat	0.0	0.33		-0.14		0.0		0.0			-0.4			0.0	
C. Grain	0.0	-0.26		0.52		0.0		0.0			0.0			0.0	
Sugar	0.0	0.0		0.0		0.10		0.0			0.0			0.0	
Dairy	0.0	0.0	0.0	-0.02	-0.01	0.0	0.0	0.06	0.06	0.0	0.01	-0.01	0.0	0.0	-0.02
R. Meat	0.0	0.0	0.05	0.0	-0.19	0.0	0.0	0.0	0.04	-0.12	0.12	0.30	0.0	0.0	-0.09
NR. Meat	0.0	0.0	-0.06	-0.02	0.0	0.0	0.0	0.0	-0.03	0.0	-0.05	0.00	0.0	0.31	0.0
The European Community															
Rice	0.20	0.0		0.0		0.0		0.0			0.0			0.0	
Wheat	0.0	0.30		-0.22		-0.02		0.0			0.0			0.0	
C. Grain	0.0	-0.22		0.40		-0.02		0.0			0.0			0.0	
Sugar	0.0	-0.02		-0.02		0.10		0.0			0.0			0.0	
Dairy	0.0	0.0	0.0	0.0	0.0	0.0	0.0	0.07	0.10	0.0	0.02	-0.03	0.0	0.0	0.0
R. Meat	0.0	0.0	0.0	0.0	-0.01	0.0	0.0	0.0	0.04	0.0	0.12	0.22	0.0	0.0	-0.16
NR. Meat	0.0	0.0	-0.22	-0.03	0.0	0.0	0.0	0.0	0.0	0.0	-0.06	-0.14	0.0	0.76	0.0

Table 9.A2. Continued

Short-run elasticity of supply with respect to the price of:

	Rice	Wheat	Coarse Grain			Sugar	Dairy			Ruminant meat			Nonruminant meat		
	t-1	t-1	t	t-1	t-2	t-1	t	t-1	t-2	t	t-1	t-2	t	t-1	t-2
Japan															
Rice	0.08	0.0		0.0		0.0		0.0			0.0			0.0	
Wheat	0.0	0.30		-0.15		0.0		0.0			0.0			0.0	
C. Grain	0.0	-0.20		0.30		0.0		0.0			0.0			0.0	
Sugar	0.0	0.0		0.0		0.10		0.0			0.0			0.0	
Dairy	0.0	0.0	0.0	-0.01	-0.02	0.0	0.0	0.05	0.30	0.0	-0.02	-0.02	0.0	0.0	0.0
R. Meat	0.0	0.0	0.01	-0.01	-0.04	0.0	0.0	0.0	0.02	-0.10	0.10	0.40	0.0	0.0	-0.05
NR. Meat	0.0	0.0	-0.05	-0.03	0.0	0.0	0.0	0.0	0.0	0.0	-0.02	0.0	0.0	0.33	0.0
The United States															
Rice	0.35	-0.09		0.0		-0.02		0.0			0.0			0.0	
Wheat	-0.02	0.45		-0.30		0.0		0.0			0.0			0.0	
C. Grain	0.0	-0.15		0.40		0.0		0.0			0.0			0.0	
Sugar	-0.01	0.0		0.0		0.07		0.0			0.0			0.0	
Dairy	0.0	0.0	-0.01	-0.01	0.0	0.0	0.07	0.02	0.08	0.03	0.03	-0.10	0.0	0.0	0.0
R. Meat	0.0	0.0	0.0	-0.02	-0.10	0.0	0.0	0.01	0.0	-0.20	0.24	0.32	0.0	0.0	-0.08
NR. Meat	0.0	0.0	-0.20	-0.01	0.0	0.0	0.0	0.0	0.0	0.0	-0.05	-0.02	0.0	0.61	0.0

Table 9.A3. Assumed Elasticities of Transmission of International Price Changes to Domestic Prices*

		Wheat		Coarse grain		Rice		Ruminant meat		Nonruminant meat		Dairy products		Sugar	
		P	C	P	C	P	C	P	C	P	C	P	C	P	C
Australia	SR	0.78	0.11	0.69	0.69	0.62	0.23	0.73	1.00	0.46	0.25	0.40	0.13	0.49	0.00
	LR	1.00	0.63	0.96	0.96	0.84	1.00	1.00	1.00	0.52	0.34	0.45	0.39	0.54	0.00
Canada	SR	0.68	0.68	1.00	1.00	0.90	0.90	0.27	0.08	0.08	0.83	0.06	0.06	0.07	0.12
	LR	1.00	1.00	1.00	1.00	0.90	0.90	0.46	0.40	0.40	0.85	0.40	0.40	0.25	0.60
EC-10	SR	0.09	0.08	0.24	0.13	0.11	0.11	0.24	0.14	0.12	0.62	0.08	0.08	0.00	0.00
	LR	0.20	0.11	0.58	0.26	0.46	0.22	0.45	0.45	0.76	0.76	0.30	0.30	0.00	0.00
EFTA	SR	0.11	0.11	0.15	0.15	1.00	0.30	0.01	0.01	0.13	0.16	0.06	0.06	0.00	0.00
	LR	0.79	0.79	1.00	1.00	1.00	0.30	0.04	0.04	0.68	0.16	0.19	0.19	0.00	0.00
Japan	SR	0.20	0.06	0.2	0.02	0.06	0.03	0.10	0.10	0.49	0.47	0.03	0.03	0.00	0.00
	LR	1.00	0.25	1.00	0.12	0.55	0.12	0.24	0.24	0.63	0.86	0.08	0.08	0.00	0.00
New Zealand	SR	0.20	0.20	0.20	0.36	0.90	0.90	0.77	0.51	0.10	0.10	1.00	1.00	0.60	0.50
	LR	0.49	0.49	0.60	0.60	0.90	0.90	0.78	0.63	0.20	0.20	1.00	1.00	0.70	0.70
Spain & Portugal	SR	0.18	0.18	0.35	0.35	0.25	0.25	0.24	0.24	0.32	0.50	0.14	0.14	0.06	0.07
	LR	1.00	1.00	0.49	0.49	0.71	0.71	0.69	0.69	1.00	0.50	0.41	0.41	0.90	1.00
United States	SR	1.00	1.00	1.00	1.00	0.82	0.71	0.60	0.21	1.00	1.00	0.07	0.06	0.10	0.10
	LR	1.00	1.00	1.00	1.00	1.00	1.00	0.61	0.53	1.00	1.00	0.36	0.18	0.48	0.48
USSR	SR	0.05	0.05	0.02	0.02	0.06	0.06	0.05	0.05	0.05	0.05	0.05	0.05	0.02	0.02
	LR	0.45	0.45	0.17	0.17	0.30	0.30	0.20	0.20	0.20	0.20	0.13	0.13	0.04	0.04
Other E. Europe	SR	0.05	0.05	0.02	0.02	0.06	0.06	0.05	0.05	0.05	0.05	0.05	0.05	0.02	0.02
	LR	0.45	0.45	0.17	0.17	0.30	0.30	0.20	0.20	0.20	0.20	0.13	0.13	0.04	0.04

Table 9.A3. Continued

		Wheat		Coarse grain		Rice		Ruminant meat		Nonruminant meat		Dairy products		Sugar	
		P	C	P	C	P	C	P	C	P	C	P	C	P	C
Egypt	SR	0.00	0.00	0.13	0.13	0.10	0.10	0.10	0.10	0.01	0.01	0.11	0.11	0.15	0.15
	LR	0.00	0.00	0.20	0.20	0.50	0.50	0.17	0.17	0.20	0.20	0.11	0.11	0.47	0.47
Nigeria	SR	0.23	0.23	0.31	0.31	0.22	0.22	0.18	0.18	0.40	0.40	0.30	0.34	0.05	0.05
	LR	0.64	0.64	0.53	0.53	0.52	0.52	0.42	0.42	0.60	0.60	0.40	0.40	0.30	0.30
South Africa	SR	0.50	0.50	0.90	0.90	0.70	0.70	0.80	0.80	0.90	0.90	0.30	0.30	0.30	0.30
	LR	1.00	1.00	1.00	1.00	1.00	1.00	0.90	0.90	0.90	0.90	0.50	0.50	0.50	0.50
Other Sub-Saharan Africa	SR	0.20	0.20	0.30	0.30	0.20	0.20	0.18	0.18	0.40	0.40	0.34	0.34	0.05	0.05
	LR	0.60	0.60	0.50	0.50	0.60	0.60	0.42	0.42	0.60	0.60	0.40	0.40	0.30	0.30
Other N. Africa & Middle East	SR	0.00	0.00	0.00	0.00	0.10	0.10	0.10	0.10	0.02	0.02	0.10	0.10	0.15	0.15
	LR	0.00	0.00	0.00	0.00	0.50	0.50	0.20	0.20	0.20	0.20	0.25	0.25	0.50	0.50
Bangladesh	SR	0.24	0.24	0.60	0.60	0.71	0.13	0.38	0.38	0.30	0.30	0.13	0.08	0.00	0.00
	LR	1.00	1.00	0.85	0.85	0.74	0.19	0.60	0.60	0.60	0.60	0.23	0.23	0.00	0.00
China	SR	0.44	0.05	0.54	0.05	0.35	0.05	0.48	0.05	0.17	0.05	0.10	0.05	0.19	0.05
	LR	0.60	0.60	0.87	0.70	0.58	0.40	0.66	0.50	0.25	0.22	0.16	0.12	0.23	0.20
India	SR	0.15	0.15	0.14	0.14	0.17	0.17	0.15	0.15	0.15	0.15	0.15	0.15	0.09	0.09
	LR	0.90	0.90	0.80	0.80	0.26	0.26	0.40	0.40	0.60	0.60	0.25	0.25	0.20	0.20
Indonesia	SR	0.09	0.09	0.47	0.46	0.20	0.05	0.05	0.05	0.05	0.20	0.05	0.02	0.02	0.02
	LR	1.00	1.00	0.94	1.00	0.60	0.40	0.40	0.40	0.40	0.40	0.20	0.20	0.20	0.20
Korea	SR	0.17	0.50	0.14	0.14	0.00	0.00	0.07	0.07	0.34	0.32	0.02	0.02	0.02	0.02
	LR	0.35	1.00	0.38	0.39	0.00	0.00	0.29	0.29	0.76	1.00	0.06	0.06	0.20	0.20

Table 9.A3. Continued

		Wheat		Coarse grain		Rice		Ruminant meat		Nonruminant meat		Dairy products		Sugar	
		P	C	P	C	P	C	P	C	P	C	P	C	P	C
Pakistan	SR	0.05	0.05	0.52	0.52	0.31	0.11	0.13	0.13	0.13	0.13	0.15	0.15	0.20	0.35
	LR	0.07	0.07	0.70	0.70	0.58	0.13	0.60	0.60	0.60	0.60	0.23	0.23	0.40	0.39
Philippines	SR	0.53	0.53	0.33	0.37	0.07	0.06	0.05	0.05	0.08	0.08	0.01	0.01	0.31	0.31
	LR	0.60	0.60	0.69	0.50	0.15	0.08	0.20	0.20	0.16	0.16	0.20	0.10	0.41	0.41
Taiwan	SR	0.09	0.42	0.40	0.91	0.24	0.22	0.54	0.08	0.43	0.20	0.01	0.01	0.51	0.51
	LR	0.60	1.00	0.43	1.00	1.00	1.00	0.93	0.62	0.53	0.32	0.20	0.20	0.73	0.73
Thailand	SR	0.40	0.40	0.85	0.85	0.49	0.31	0.17	0.17	0.18	0.18	0.01	0.01	0.24	0.24
	LR	0.60	0.60	1.00	1.00	0.74	0.58	0.30	0.30	0.50	0.50	0.20	0.20	1.00	1.00
Other Asia	SR	0.05	0.05	0.40	0.30	0.20	0.15	0.15	0.20	0.32	0.20	0.00	0.20	0.20	0.20
	LR	0.20	0.20	0.80	0.50	0.80	0.50	0.60	0.20	0.60	0.20	0.00	0.20	0.40	0.20
Argentina	SR	0.80	0.80	0.70	0.70	0.56	0.56	0.58	0.77	0.43	0.66	0.34	0.34	0.00	0.00
	LR	1.00	1.00	0.80	0.80	0.56	0.56	0.63	0.90	0.46	0.80	0.35	0.35	0.00	0.00
Brazil	SR	0.42	0.42	0.57	0.35	0.16	0.26	0.44	0.44	0.72	0.72	0.54	0.54	0.24	0.24
	LR	0.79	0.79	1.00	0.42	0.46	0.32	0.60	0.60	0.77	0.77	0.54	0.54	0.90	0.90
Cuba	SR	0.00	0.00	0.11	0.11	0.02	0.02	0.00	0.00	0.24	0.24	0.00	0.00	0.00	0.00
	LR	0.00	0.00	0.30	0.30	0.20	0.20	0.00	0.00	0.24	0.24	0.00	0.00	0.00	0.00
Mexico	SR	0.25	0.25	0.31	0.21	0.37	0.37	0.13	0.13	0.50	0.50	0.10	0.10	0.00	0.00
	LR	0.61	0.61	1.00	0.23	0.47	0.47	0.34	0.34	0.50	0.50	0.20	0.20	0.00	0.00
Other Latin	SR	0.50	0.50	0.50	0.50	0.60	0.60	0.50	0.50	0.50	0.60	0.02	0.20	0.00	0.00
America	LR	0.80	0.80	0.80	0.80	0.80	0.80	0.80	0.80	0.80	0.90	0.20	0.30	0.00	0.00

*SR and LR refer to short run and long run elasticities (with a Nerlovian geometric lag structure connecting them); P and C refer to domestic producer and consumer prices, respectively.

REFERENCES

Abbott, P. C. "Modelling International Grain Trade With Government-Controlled Markets," American Journal of Agricultural Economics 61(No. 1, 1979):22-31, February.

_________. (1988). "Estimating U.S. Agricultural Export Demand Elasticities: Econometric and Economic Issues," Ch. 3 in this volume.

Anderson, K. and R. Tyers. (1987). "Global Effects of Liberalizing Trade in Agriculture," Thames Essay No. , London: Trade Policy Research Centre.

Binkley, J. K. and L. McKinzie. "Alternative Methods of Estimating Export Demand: A Monte Carlo Comparison," Canadian Journal of Agricultural Economics 29(1981):189-202.

Bredahl, M. E., W. H. Meyers, and K. J. Collins. "The Elasticity of Foreign Demand for US Agricultural Products: The Importance of the Price Transmission Elasticity," American Journal of Agricultural Economics 61(No. 1, February 1979):58-63.

Gardiner, W. H. and P. M. Dixit. "Price Elasticity of Export Demand: Concept and Estimates," FAER No. 228, U.S. Department of Agriculture, Washington, D.C. February 1987.

Johnson, P. R. "The Elasticity of Foreign Demand for U.S. Agricultural Products," American Journal of Agricultural Economics 59(No. 4, 1977):735-36, November

Johnson, S. R., A. W. Womack, W. H. Meyers, R. E. Young II and J. Brandt. (1985). "Options for the 1985 Farm Bill: An Analysis and Evaluation," FAPRI Report No. 1-85, Food and Agricultural Policy Research Institute, University of Missouri - Columbia and Iowa State University.

Meyers, W. H., S. Devadoss, and M. D. Helmar. "Agricultural Trade Liberalization: Cross-Commodity and Cross-Country Impacts Products," Journal of Policy Modeling 9(No. 3, 1987):455-84.

Meyers, W. H. and M. D. Helmar. "Trade Implications of the Food Security Act of 1985," Staff Report No. 86-SR4, Centre for Agricultural and Rural Development, Iowa State University, February 1986.

Orcutt, G. H. "Measurement of Price Elasticities in International Trade," Review of Economics and Statistics 32(No 2, 1950): 113-32, May.

Tyers, R. "Agricultural Protection and Market Insulation: Analysis of Interntional Impacts by Stochastic Simulation," Research Paper No. 111, Australia-Japan Research Centre, Canberra, May 1984.

_______. "International Impacts of Protection: Model Structure and Results for EC Agricultural Policy," Journal of Policy Modelling 7(No 2, 1985):219-51.

Tyers, R. and K. Anderson. Distortions in World Food Markets, Cambridge: Cambridge University Press, forthcoming.

Chapter 10

Discussion: William H. Meyers, Andrew Schmitz, and Robert Thompson

Concluding Comments: Alex F. McCalla

This chapter provides a brief summary of the main contributions of the individual papers and indicates where the agricultural economics profession is left on the subject of elasticities in international trade modelling.

DISCUSSION

William H. Meyers

This has been an excellent session and I congratulate both the planners and the authors for making it successful. I would like to comment on and amplify some of the points that were made. First of all in the area of estimation and, secondly, with respect to the uses of these estimates.

The Thursby's paper presented a method of estimation as close to value free as I have seen, and it is very appealing. Phil Abbott provided a very complete treatment of estimation problems and procedures that will be a useful reference for those engaged in this kind of work. I agree with his rejection of the direct estimation of U.S. export demand as a means of identifying the export response behavior. However, I feel it is important to recognize the reasons why that method has been used so much in the past. In early U.S. commodity models, exports were often exogenous variables. This became unacceptable in the 1970s as exports became more and more important and more and more a source of variability in demand and prices. Commodity modelers went first to single equation specifications of export demand as a way of endogenizing exports without adding a large, cumbersome foreign market component to what were relatively compact domestic commodity models.

In the Food and Agriculture Policy Research Institute we have resolved this issue by directly modeling individual countries or regional supply and demand behavior in the context of a world nonspatial equilibrium model. If, then, we need a single equation model to operate the U.S. model by itself, we use the derived reduced form based on the detailed world model. This derived reduced form has only current and lag prices as endogenous variables; and all exogenous variables are combined in the constant term. Thus, this reduced form must be derived for different sets of exogenous assumptions, but it then closely replicates behavior of the trade model with respect to prices and exports.

Kym Anderson's paper, aside from documenting the behavior underlying his model, pointed out some important problems in estimating the price response of exports. These problems have to do with the variability of the response of elasticity with respect to the year in which it is computed and with respect to the assumptions about price transmission elasticity. As an example, the computed long-run export response to price changes for coarse grains has an elasticity of -1.3 for the period 1980-1982, but an elasticity of -0.4 at 1988 levels of price and quantities. This kind of variability should make us cautions about making absolute statements on what the export response elasticity is. The degree to which the export response elasticity is influenced by the price transmission elasticities underlines once more the importance of more research on the price transmission linkages. These equations need to be more completely specified and need to include variables that reflect changes in policy over the time period of the estimation and the effect of changing market conditions on these policies.

With respect to uses of export demand elasticities, I would like to underline Jim Houck's comment that the elasticity is a variable and not a parameter. It is important to remind ourselves and to communicate with the outside world that the elasticities that we present in a particular paper or study do not necessarily apply in general. Most of the components required to compute the export response elasticity are themselves variables that change over time. This is frequently overlooked even in discussions among professions and much more so when these discussions occur by users outside the economics profession.

These qualifications lead me to believe that the debate over agricultural policy directions has focused too narrowly on this one, ill-defined and highly controversial variable, to the exclusion of broader issues. Did we really mean to argue that if we could design a policy that increased the value of exports that we have achieved the primary objective in agricultural policy? If we had strong empirical evidence that export demand was inelastic, would that have led us to strongly argue for a high price policy? How much of our arguments in the policy debate were based on empirical evidence and how much on our economic beliefs or philosophies? I am reminded of the article, "The Rhetoric of Economics" by McCloskey in the Journal of Economic Literature (June 1983) in which he argued that economists don't depend on scientific testing nearly as much as we claim and that we would be better off if we admitted that to ourselves and to others. I would argue that the elusiveness of the "export demand elasticity" is not a failure of research by the profession but it is inherent in nature of the elasticity itself. This should lead us not to berate ourselves for inadequate research on the subject, but rather to be extremely cautious in drawing policy implications from what we know to be a fickle variable.

Andrew Schmitz

There are several conceptual problems associated with estimating trade elasticities and I would like to briefly mention a few in particular. It is conceptually difficult to work with importers' excess demand curves in today's trading environment. What are the import demands facing U.S. agriculture? If all the agricultural trade distortions were taken into account it would be found that the theoretical construction of excess demand relationships becomes extremely difficult. How can an economist estimate a demand relationship which may be impossible to derive theoretically?

Institutional arrangements are often overlooked by model builders. The importance of this omission cannot be overstated. Given the institutional arrangements which exist in the world wheat market it is difficult to imagine how the demand for U.S. wheat exports can be elastic. Since the 1985 Farm Bill was made law, U.S. wheat export prices have fallen, U.S. export volume has increased, but the value of U.S. exports has not increased. This implies the demand for U.S. exports is inelastic. Institutional arrangements prevent the U.S. from operating on the elastic portion of the importer's demand curve.

A basic problem with the 1985 Farm Bill is that it did not anticipate the impact lower wheat prices would have on competing exporters such as Canada and the EC. The Farm Bill may have been an attempt to force out competition (i.e., shift supply) rather than directly move along the excess demand curve although eventually a reduced output by competitors shifts the excess demand curve facing the U.S. Competing exporters did not reduce supplies significantly at lower world prices and this is a further illustration of the importance of incorporating institutional behavior into trade models. This behavior makes the excess demand curve facing the U.S. relatively price inelastic.

The drought in North America in 1988 perhaps will make the demand appear like it is price elastic. However, a North American drought causes excess supply to shift leftward. This can cause the total value of U.S. exports to rise. Clearly, in this case, the excess demand curve can still be highly price inelastic where the increase in total value comes about because of reduced quantities.

Robert Thompson

During the 1985 Farm Bill debate the price responsiveness of import demand for U.S. farm products was the single most important issue. It was ultimately determined that the U.S. faces an elastic import demand.

I disagree with Willy Meyers' comment that the import demand elasticity is a variable rather than a parameter. There must be a single indicator of price responsiveness.

There are several issues which have been brought out by this Symposium:

- A single-country excess demand schedule cannot be properly estimated. A multi-country framework must be utilized.
- A general equilibrium approach is preferable.
- Differentiated products must be incorporated in the model.
- Static results will differ from dynamic behavior.
- Price response is asymmetrical. The import response for a price increase may be greater than that for a price decrease.
- Stochastic models are more useful than deterministic models.

- The recent trade literature on imperfect competition should be incorporated into modelling efforts. More political-economy modelling is required. The policy content of existing models is inadequate.

CONCLUDING COMMENTS

Alex F. McCalla

This Symposium has addressed an important, but obviously complex issue of critical importance to trade analysis. At this point it might be appropriate to ask a difficult question--do we know more about this topic than when we came to the Symposium? The answer is, as is usual in economics--it depends. If we expected to discover a consensus on the "correct" elasticity value, the answer is a resounding--NO!! In fact, we may have been disabused of even believing there is a single number. If we expected a clear statement of how to measure elasticities so that we could generate the true number, again the answer is NO. We have heard that direct measures are biased downwards, while synthetic or calculated methods are biased upwards, but we don't know where in the middle the right estimates are.

If we came expecting a clear outline of the conceptual form of the elasticity of demand for exports, we were also disappointed. In fact, we may have discovered that we are not always clear on what elasticity we are discussing. In this conference, I have heard four conceptually different elasticities discussed, often without the recognition that they are different. These are:

1. the elasticity of Japanese demand for U.S. wheat
2. the elasticity of world demand for wheat (presumably the horizontal summation of all importer excess demand functions)
3. the elasticity of demand for U.S. wheat exports (presumably #2 minus the summation of excess supplies for competing exporters)
4. the elasticity of demand for U.S. wheat (which includes, in addition to #3, the domestic demand for wheat).

A priori we would expect all of these to be different and clearly the measurement problems are different. Even in the concluding debate, the argument focused, I think, on the third one when in fact for U.S. policy purposes, #4 seems to be the more appropriate one.

What then, have we learned from the papers presented at this Symposium? First, we have learned that the problem is not simple. Rather, we should have learned how incredibly complex the issue is. In fact, there is a real question as to whether seeking a "single number" is empirically possible or conceptually defensible. Out of this exposure, we should be more cautious about claims and counter claims about who is "right". Understanding the limitations of what we know is the first step in pushing back the frontier of knowledge. It also should contribute to us being less sure but much wiser. If we gained that wisdom, then the Symposium has been a success.

Contributors

Philip C. Abbott
Dept. of Ag. Econ.
Purdue University
West Lafayette, IN 47907

Irma Adelman
Dept. of Ag. &
Resource Econ.
University of California
Berkeley, CA 94720

Kym Anderson
Dept. of Econ.
University of Adelaide
Adelaide, South Australia
AUSTRALIA

David Blandford
Dept. of Ag. Econ.
Cornell University
Ithaca, NY 14853

Christine Bolling
USDA/ERS/IED
1301 New York Avenue, N.W.
Washington, D.C. 20005-4788

Colin A. Carter
Dept. of Ag. Econ.
University of California
Davis, CA 95616

John Dutton
Dept. of Econ. and Business
North Carolina State University
Raleigh, NC 27695

Walter H. Gardiner
USDA/ERS
1301 New York Avenue, N.W.
Washington, D.C. 20005-4788

Ellen W. Goddard
School of Ag. Econ. &
Ext. Educ.
University of Guelph
Guelph, Ontario
CANADA N1G 241

Thomas Grennes
Dept. of Econ. and Business
North Carolina State University
Raleigh, NC 27607

Alex F. McCalla
Dept. of Ag. Econ.
University of California
Davis, CA 95616

William H. Meyers
Dept. of Econ.
Iowa State University
Ames, IA 50011

Sherman Robinson
Dept. of Ag. &
Resource Econ.
University of California
Berkeley, CA 94720

Andrew Schmitz
Dept. of Ag. &
Resource Econ.
University of California
Berkeley, CA 94720

Robert Thompson
School of Agriculture
Purdue University
West Lafayette, IN 47907

Jerry G. Thursby
Dept. of Econ.
Purdue University
West Lafayette, IN 47907

Marie C. Thursby
National Bureau of Economic Research
Dept. of Econ.
Purdue University
West Lafayette, IN 47907

Rod Tyers
Dept. of Econ.
University of Adelaide
Adelaide, South Australia
AUSTRALIA

Index

Abbott, P. C., 9, 70, 72, 260,262, 297
Aggregation, 8-9, 59-61,88-92, 99, 104
  over countries, 90-92, 94, 117
  over currencies, 91-92
  regional, 59-60
  sectoral, 143
  by world models, 63-64
Agricultural Outlook index, 102
Akaike, H., 27
"Almost ideal demand system," 6
Amemiya, T., 27
"Analytical least squares," 59, 61, 62
Anderson, Kym, 258, 298
Appreciation, 120(table)
Argentina, 102, 103, 187(table), 231, 236, 237, 248(table), 250(table)
  demand elasticities, 81(table), 90, 122(table), 241, 247(table), 251(table), 259, 263, 264-266(tables)
  exchange rates, 168
  farm policy, 170
  farm prices, 3, 166, 167, 168, 169, 178, 180(figure), 181(figure), 184(figure), 185(table), 188-192(tables)
  price elasticities, 294(table)
Armington, P. S., 5-6, 67-68, 235, 236
Armington models, 67-69, 76-77, 79, 80, 82(table)
Askari, H., 75
Australia, 3, 102, 122(table), 231, 232, 236, 237, 238, 240, 241, 247(table), 248(table), 249(table), 251(table), 259, 264-266(tables, 292(table)

Balance of trade. See Trade, balance
Bangladesh, 293(table)
Batten, Dallas, 106
Becketti, Sean, 102
Beef
  demand elasticities, 226, 231-253
  markets, 231-233, 234, 241-242, 243, 247-251(tables)
  price elasticities, 240, 247(table), 249-250(table)

production systems, 232, 234
supply, 233-234, 246(table), 248(table)
See also Meat
Behavioral models, 62-63, 64, 65-66, 79, 80
Belgium, 101
Belongia, Michael T., 101, 106
Bias, 8, 47(n10), 56-61, 115, 116. See also Estimation, errors
Binkley, J. K., 8, 59, 255
Blandford, D., 71, 72
Bolivia, 166, 168-169, 178, 180(figure), 181(figure), 184(figure), 185(table), 187-192(tables)
Boot, ______, 115
Brazil, 101, 103, 187(table), 237, 248(table), 250(table)
demand elasticities, 81(table), 241, 247(table), 251(table), 260, 264-266(tables)
exchange rates, 168
farm policies, 170-171
farm prices, 166, 168, 169, 178, 180(figure), 181(figure), 184(figure), 185(table), 188-192(tables)
price elasticities, 294(table)
Bredahl, M. E., 9, 69-70, 78, 163, 227, 260
Bretton Woods system, 114
Breusch, T. S., 29
Burgess, D. F., 22, 23
Burris, A. E., 7, 67
Buse, R. C., 229
Byerlee, D., 70
CAIRNS group, 4
Canada, 3, 30, 31, 38(table), 39(table), 77, 78, 101, 102, 231, 234, 237, 238, 242, 248(table), 249(table), 300
export demand, 81(table), 122(table), 241, 242, 247(table), 251(table), 259, 264-266(tables)
import demand, 38(table), 39(table)
price elasticities, 288-289(table), 292(table)
supply elasticities, 290(table)
Capel, R. E. 201
Cargill, 91
Cartels, 91
Carter, Jimmy, 145
CES. See Constant elasticity of substitution function
CGE. See Computable general equilibrium model
Chamberlain, G., 28
Chambers, R. G., 7, 105-106
Chile, 187(table)
farm policies, 174-175, 177
farm prices, 167, 168, 169, 177, 178, 180(figure), 182(figure), 185(table), 188(table), 189(table), 191(table), 192(table)
Clements, K. W., 21
Coarse grains
demand elasticities, 54, 60, 78, 79, 81(table), 261, 264-267(tables)
markets, 259
price elasticities, 97, 121(table), 298

price transmission, 256, 288-289(table), 292-294(table)
supply elasticities, 290-291
Cobb-Douglas functions, 65
Collins, H. Christine, 177, 178
Collins, K. J., 6, 9, 69-70, 78, 163, 227, 260
Colombia, 187(table)
exchange rates, 168
farm prices, 166, 168-169, 175, 178, 180(figure), 182(figure), 184(figure), 185(table), 188-192(tables)
Commodity aggregation. See Aggregation
Commodity markets
share, 53
structure of, 2
substitutability and, 54-55
Computable general equilibrium (CGE) model, 138-144
Constant elasticity of substitution (CES) function, 65, 68, 69, 235, 236
Constrained estimation, 61-66
Conway, Roger K., 7
Cooley, T., 28
Corlett, ________, 8
Corn
demand elasticities, 120(table), 123(tables)
markets, 199, 207(table), 209(table), 211(table), 212(table), 214(table, 216(table), 218(table), 220(table), 222-223(tables)
price elasticities, 123(tables), 126-127(table), 128(table), 166, 167, 168-169, 181-183(figure)
price transmission, 178, 185-186(table), 189(table), 191(table), 192(table)
repeat purchase of, 199
supply elasticities, 97, 120(table)
Costa Rica, 169, 182(figure), 185(table), 187(table), 189(table), 191(table), 192(table)
Cotton, 95
Cox, D. R. 27, 28
Cox, Michael, 101
Cuba, 294(table)
Cummings, J. T., 75
Currencies. See Aggregation, over currencies; Devaluation

Dairy products
export demand, 259, 264-266(tables)
price transmission, 256, 259, 288-289(table), 292-294(table)
supply elasticities, 290-291(table)
Davidson, R., 29
Deaton, A. S. 28
Delphi process, 66, 80. See also Estimation, synthetic
Demand response, 2, 99
Depreciation
currency, 104, 106, 107, 123(tables)

of dollar, 92
Devaluations, 95, 99, 100, 187(table)
of dollar, 101
trade balance and, 2, 7
Dixit, P. M., 21, 54-55, 56, 78, 82(table), 255
Dominican Republic, 168, 187(table), 189(table), 191(table), 192(table)
Dornbusch, Rudiger, 100
Drabenstott, Mark, 106
Dutton, John , 112

EC. See European Community
Ecuador, 174, 187(table), 189(table)
Edwards, Sebastian, 103
EFTA, 292(table)
Egypt, 81(table), 199, 201, 203, 204, 207-223(tables), 293(table)
El Salvador, 168, 169, 182(figure), 185(table), 187(table), 189(table)
Estimation, 7-10, 53-85, 98-99, 115-118
constrained, 61-66
errors, 56-61, 115, 116, 119, 237, 238
by OLS, 237-238
synthetic, 55-56, 59, 62, 65, 66, 75-76, 77, 79, 80, 82(table), 99, 255
variables, 129-130
European Community (EC), 4, 96, 100, 231, 232, 237
export demand, 81(table), 122(table), 241, 247(table), 251(table), 259, 260, 264-266(tables)
export subsidy by, 4, 94
price elasticities, 250(table), 288-289(table), 292(table)
supply elasticities, 248(table), 290(table)
U.S. trade relations, 198, 199, 202, 203, 205, 207-223(tables), 300
European Monetary Union, 100
Exchange rates
effects of, 100, 101, 104-106, 107, 118-119, 163-164
export quantity and, 106-107, 124(table), 125(table)
indices, 92, 100-104, 106, 119
prices and, 104-106, 128(table), 168
real, 103-104, 105, 118, 150
theoretical issues, 108-114
trade models and, 7, 87-135
trade value and, 107
transmission, 178, 188-190(tables)
Export demand, 106-107, 301-302
behavioral parameters, 62-63, 64, 65-66, 79, 80, 94, 163-194
constrained estimation, 61-66
direct estimation, 55, 61, 65-66, 79, 82(table)
domestic policy and, 95-98
elasticities, 33-35, 47(n7), 54-66, 75-79, 90-92, 93-94, 118, 225-253
estimation of, 7-10, 53-85, 98-99, 115-119, 129-130, 237, 238, 255

structural change in, 73-74
synthetic estimation, 55-56, 59, 62, 65, 66, 75-76, 77, 79, 80, 82(table), 99, 255
technological change and, 117
Export subsidies. See Subsidies
Export supply, 98, 105

Factor markets, 99
FAPRI. See Food and Agricultural Policy Research Institute
Federal Republic of Germany, 30, 31, 38(table), 39(table), 101
Feed demand, 62, 65, 73-74, 80
Figueroa, E. E., 196, 201
Financial markets, 99
Floyd, J. B., 9
Food and Agricultural Policy Research Institute (FAPRI), 3, 89, 97, 298
Food crisis, 3
Food security, 73
Forecasting models. See Modeling
France, 3, 100, 101, 102

Game theory, 2
Garcia, __________, 172
Gardiner, W. H., 21, 54-55, 56, 78, 82(table), 255
General Agreement on Tariffs and Trade (GATT), 4
Generalized Box-Cox form, 235-236
Ginsburgh, Victor A., 115
GLS model, 255-295
Goddard, E. W., 196, 238
Godfrey, L. G., 29
Goldfeld, S. M., 29
Goldstein, Morris, 7, 93, 98
Goodloe, C. A. 71, 73
Goodness of fit procedures, 27
Goods. See Commodity markets; Inputs; Product homogeneity
Government intervention, 9, 55, 117, 118, 263
in domestic pricing, 257
export demand estimation and, 63
inventory demand and, 96-97
in Latin America, 167-168, 169-175, 178
in LDCs, 70, 73
monetary, 100
policy regimes, 71, 73
price transmission and, 69-71, 93-94, 95, 163, 260
time lags and, 93
Grains. See Coarse grains; Wheat
Gregory, R., 22
Grennes, T., 6, 60, 67, 112
Griliches, Z., 24
Group of Ten, 101, 103
Guatemala, 167, 168, 169, 182(figure), 185(table), 187(table), 189(table), 191(table), 192(table)

Hakkio, Craig, 102
Hall, A. D., 29
Harberger, A. C., 7
Hausman, J. A., 30
Heady, E. O., 77, 82(table)
Helmar, M. D., 261
Henneberry, David, 106

Henneberry, Shida, 106
Hertel, Thomas W., 142, 143
Holland, F. D., 63, 79
Honduras, 118, 182(figure), 186(table), 187(table), 189(table), 191(table), 192(table)
Hong Kong, 101
Honma, M., 77, 82(table)
Horner, F. B., 225, 226, 227
Houck, Jim, 298

Identification error, 57-58. See also Estimation, errors
Imperfect substitutes model, 5, 21-24
Import demand, 30, 31, 37(table), 38(table), 47(n9), 300-301
Income
- elasticities, 32, 34, 35, 39(table)
- indices, 155(table)
- per capita, 117

India, 81(table), 293(table)
Indonesia, 81(table), 293(table)
Inflation
- exchange rage indices and, 102-103
- price transmission and, 167

Inputs, 23
Institutional behavior, 299-300
Intercambio Ltd., 6
Inventory demand, 96-97. See also Export demand
Italy, 101

Japan, 30, 31, 33, 101, 198, 207-223(tables), 232, 237, 242, 248(table)
- demand elasticities, 81(table), 241, 247(table), 263, 264-266(tables, 301
- import demand, 38(table), 39(table), 259
- import supply, 259, 260
- markets, 199, 207-223(tables), 231, 234
- price elasticities, 32, 201, 205, 240, 288-289(table), 292(table)
- supply elasticities, 248(table), 291(table)

Japanese Food Agency (JFA), 33
Jarvis, L. S., 233-234
JFA. See Japanese Food Agency
Johnson, B. F., 73
Johnson, P. R., 6, 9, 60, 67, 78, 163, 261
Johnson, S. R., 261
Junz, H. B., 6
Just, Richard E., 7, 105-106

Kahn, Moshin S., 7, 93, 98
Kohli, U. R., 22, 23-24
Kolstad, C. D., 67
Kost, W. E., 7
Koyck, L. M., 7
Krissoff, Barry, 92, 95

Labys, W. C., 7
Lattimore, R., 71
"Law of one price," 2, 47(n5), 166
LDCs. See Less developed countries
Leamer, E. E., 7, 28, 202
Lee, T. C., 198

Leonard, H., 28
Leontief functions, 65
LES. See Linear expenditure system
Less developed countries (LDCs)
  diet composition in, 73-74
  feed demand in, 62
  government intervention in, 70, 73
  prices in, 70
Linear expenditure system (LES), 64, 65, 76, 80
Liu, K., 70, 75
Livestock markets, 233. See also Beef
LOME agreement, 241
Longmire, Jim, 89

McAleer, M., 28
McCloskey,__________, 299
Machlup, F., 7
MacKinnon, J. G., 28, 29
McKinzie, L., 8, 59, 255
Macro shocks, 144-147
Malaysia, 101
Mallows, C. L., 27
Manufactures, 98
Markets
  insulation of, 260-261, 263
  interdependency, 99, 100
  models, 3-8, 64-65, 74-75, 298, 299
  share, 196-224
  structure of, 2, 91
  See also Commodity markets
Marketing strategy, 242-243
Market share model, 203-205, 218-223(tables)
Markov model, 197-200, 204, 205, 209-211(tables)
Marshall-Lerner condition, 2
Meat
  demand elasticities, 73, 259, 264-266(tables)
  price transmission, 256, 288-289(table), 292-294(table)
  supply elasticities, 65, 235, 290-291(table)
  See also Beef
Meilke, Karl, 89, 97
Mellor, J. W., 73
MERM index, 106
Mexico, 81(table), 101, 103, 187(table), 198, 201, 203, 207-223(tables), 237, 248(table)
  demand elasticities, 247(table)
  farm policies, 173-174
  farm prices, 166, 167-168, 169, 178, 180(figure), 183(figure), 184(figure), 186(table), 188-192(tables)
  markets, 199, 207-223(tables)
  price elasticities, 294(table)
Meyers, W. H., 9, 69-70, 78, 163, 227, 260, 261, 300
Michigan State model, 97
Modeling, 2, 3-8, 64-65, 74-75, 298, 299
Money supply, 114
Morey, Art, 89, 92, 95
Morgan,__________, 8
Morgan Guaranty Trust, 101, 106
Multicollinearity, 58-59, 60, 64, 89

Nerlove, M., 7, 74-75

Netherlands, 101
New Zealand, 231, 236, 237, 240, 241, 242, 247-249(tables), 251(table), 259, 263, 264-266(tables), 292(table)
Nicaragua, 167, 168, 169, 183(figure), 186(table), 189(table)
Nigeria, 81(table), 293(table)
Non-nested procedures, 27-28

Oceania, 81(table), 231
OECD. See Organization for Economic Cooperation and Development
OLS. See Ordinary least squares
Orcutt, G. H., 7, 8, 57, 98, 255
Ordinary least squares (OLS), 8, 237-238
Organization for Economic Cooperation and Development (OECD), 103
Ott, Mack, 101

Paarlberg, P. L., 72, 91
Pagan, A. R., 29
Pakistan, 294(table)
Paraguay, 166, 167, 168, 169, 177, 178, 181(figure), 183(figure), 184(figure), 186-192(tables)
Parameter sensitivity, 77-79
Partial elasticity, 99-100
Patterson, P., 60, 75, 77, 79, 82(tables)
People's Republic of China, 196, 198, 199, 201, 202, 207-223(tables)
  demand elasticities, 81(table), 264-266(tables)
  import supply, 259
  price elasticities, 293(table)
Perfect substitutes model, 5, 20-21, 47(n5)
Peron, Juan, 170
Peru, 187(table)
  exchange rates, 168, 172
  farm policies, 171-172
  farm prices, 166, 168-169, 181(figure), 183(figure), 186(table), 188(table), 189(table), 191(table), 192(table)
Pesaran, M. H., 28
Philippines, 101, 294(table)
Policy analysis, 2
Portugal, 292(table)
Price
  behavior, 163-194
  depreciation and, 104-105, 106
  domestic, 69, 70-71, 94-95, 106
  elasticities, 1-4, 5-6, 8-9, 34, 35, 75, 88-89, 91, 93-96, 98, 121(table), 255
  exchange rates and, 104-106, 128(table), 168
  government intervention and, 69-71, 93-94, 95, 163, 260
  indices, 155(table), 157(table)
  producer, 157(table)
  supports. See Subsidies

transmission of, 69-71, 79, 80, 91, 93-95, 96-97, 118, 163-169, 176-179, 185-186(table), 189-192(tables), 255-295, 298
world, 69-71, 75, 96, 164
See also Estimation
Product categories, 88-89, 98
Product homogeneity, 226, 228, 229, 234, 242, 262
Production. See Supply, response

Quandt, R., 29

Ramsey, J. B., 29
Rawleigh Foundation, 2
Reagan, Ronald, 137, 145, 147
Regional aggregation, 59-60
Repeat purchase, 198
Republic of Korea, 101, 199, 203, 207-223(tables), 232, 237, 293(table)
RESET procedure, 29-30, 31, 32, 48(n13)
Restitution, 96. See also Subsidies
"Rhetoric of Economics, The" (McCloskey), 299
Rhomberg, R. R., 6
Rice
demand elasticities, 264-267(tables)
markets, 259
price transmission, 256, 288-289(table), 292-294(table)
supply elasticities, 290-291(table)
Richardson, J. D., 237
Rigaux,____________, 201
Roe, Terry, 9, 91
Roningen, V. O., 70, 75
ROW, 81(table)

Sain, G., 70
Sarris, A. H., 8
Schmidt, P., 29
Schmitz, Andrew, 8, 91
Schuh, G. E., 7
Schwartz, M., 7
Shane, Matthew, 9, 91
Sharples, J. A., 62-63, 69, 71, 73, 79, 80, 91
Simultaneous equation bias, 8
Singapore, 101
Smith, Adam, 91
Smuggling, 95
South Africa, 3, 81(table)
Soviet Union, 77-78, 81(table), 231, 237, 259
demand elasticities, 264-266(tables)
import supply, 259
price elasticities, 292(table)
trade elasticities, 263
U.S. trade relations, 196-197, 199-200, 203, 204, 207-223(tables)
Soybeans
demand elasticities, 54, 120(table), 123(tables)
price elasticities, 94, 95, 97, 121(table), 123(tables), 126-127(table), 128(table), 166, 167, 169, 184(figure)
price transmission, 178, 185-186(table), 190-192(tables)
supply elasticitiies, 97, 120(table)
Spain, 292(table)
Spatial equilibrium, 2

Spatial equilibrium model, 63-64, 79, 82(table)
Special Drawing Rights, 101, 106
Specification, 24-30
  bias, 47(n10)
  error, 24-25, 29-30, 41-46, 48(n15), 56-57
  searches, 28-29
Spriggs, J., 71
Stability. See Food security
Stern, R. M., 7, 31, 202
Stock demand. See Inventory demand
Stock management, 62-63, 71-73, 74, 79
Stockpiling, 72
Structured models, 64-65
Subsidies, 53-54, 94, 96, 97-98, 107. See also Government intervention
Substitution
  elasticities, 54-55, 65, 68, 79, 82(table), 200-203, 205, 212-217(tables), 235, 243
  imperfect, 5, 21-24, 226-230, 239, 243
  perfect, 5, 20-21, 47(n5), 195-196, 225, 230, 239, 242, 243
Sugar, 95
  demand elasticities, 264-266(tables)
  markets, 260
  price transmission, 256, 288-289(table), 292-294(table)
  supply elasticities, 290-291(table)
Supply
  irreversibilities, 75
  modeling, 74-75
  response, 74-75, 92, 95-98
Sweden, 101

Taiwan, 101, 203, 207-223(tables), 294(table)
Taplin, J. E. H., 225
Tariff equivalents. 70-71
Tariffs, 2
Taylor, C. 74
Technology, 117
Thailand, 3, 81(table), 259, 263, 264-266(tables), 294(table)
Theil, H., 8, 24, 27
Thompson, R. L., 8
Thursby, J. G., 28, 29, 30
Thursby, M. C., 6, 28, 30, 60, 67, 297
Time lags, 6-7, 92-93, 99, 118
Time series analysis, 61
Trade
  balance, 2, 98, 107
  barriers, 226
  equations, 17-18, 19-30. See also Modeling
  flows, 17-52, 67-69, 71, 77
Trade Embargo Study, 66, 75-76, 77, 78, 79, 81(table), 82(table)
Translog system, 64, 65
Transportation charges, 79
Tsingas, Marison E., 142, 143
Tweeten, Luther, 8, 9, 163
Tyers, R., 258

United Kingdom, 30, 31, 32, 38(table), 39(table), 101
United States, 30, 31, 237, 240, 242, 248(table), 249(table), 259
  debt, 146

deficit, 146, 150
Department of Agriculture index, 102, 106
equilibrium analysis of, 137-162
export demand, 3, 4, 33, 53-85, 89-90, 95, 106, 108, 118, 122(table), 124(table), 125(table), 196-224, 241, 247(table), 251(table), 259, 263, 264-266(tables), 301
export supply 98, 291(table)
export transformation, 152(table)
export value, 299-300
Farm Bill (1985), 299, 300
farm policy, 3-4, 53-54, 107, 163, 179
Federal Reserve Board Index, 101, 106
Food Security Act, 3
GNP, 153(table), 154(table)
import demand, 30, 31, 38(table), 39(table)
import substitution, 152(table)
land values in, 146-147
macro policy, 137-138, 147, 150
price elasticities, 288-289(table), 292(table)
sectoral analysis, 152(table), 156(table), 158(table)
social accounting matrix, 143, 151(table)
supply elasticities, 291(table)
trade balance, 145
trade shares, 152(table)
Uruguay, 187(table), 237, 248(table), 250(table)
demand elasticities, 241, 247(table), 251(table)
farm prices, 166, 167, 177, 183(figure), 186(table), 188(table), 189(table), 191(table), 192(table)
Utility maximization, 67-68

Venezuela, 81(table), 187(table)
farm policies, 172-173
farm prices, 169, 183(figure), 186(table), 189(table), 191(table), 192(table)
Vo, De Huu, 9, 91

Wells, Gary, 6
Wheat
demand elasticities, 60, 62, 71, 78, 79, 81(table), 89-90, 106-107, 120(table), 122(table), 123(tables), 264-267(tables)
domestic supply, 97
export elasticities, 40(table), 54, 299, 301
markets, 6, 199, 200, 207(table), 210(table), 211(table), 213(table), 215(table), 217(table), 219(table), 221(table), 222-223(tables), 259
price elasticities, 92, 94, 96, 97, 99, 107, 121(table), 123(tables), 126-127(table), 128(table), 166, 168, 180-181(figure), 204

price transmission, 177, 185-186(table), 191(table), 192(table), 256, 259, 288-289(table), 292-294(table)
repeat purchase of, 199, 200
substitutability, 195-196
supply elasticities, 91, 95-96, 290-291(table)
Winters, L. A., 6
Wisconsin Tariff Commission, 2
Witherell, W. H., 7
Wool, 95
World Bank model, 97

Zwart, A. C., 71, 72

For Product Safety Concerns and Information please contact our EU representative GPSR@taylorandfrancis.com Taylor & Francis Verlag GmbH, Kaufingerstraße 24, 80331 München, Germany